BILLION
DOLLAR
BURGER

BILLION DOLLAR BURGER

Inside Big Tech's Race
for the Future of Food

CHASE PURDY

PORTFOLIO / PENGUIN

Portfolio / Penguin
An imprint of Penguin Random House LLC
penguinrandomhouse.com

Most Portfolio books are available at a discount when purchased in quantity for sales promotions or corporate use. Special editions, which include personalized covers, excerpts, and corporate imprints, can be created when purchased in large quantities. For more information, please call (212) 572- 2232 or e-mail specialmarkets@penguinrandomhouse .com. Your local bookstore can also assist with discounted bulk purchases using the Penguin Random House corporate Business-to-Business program. For assistance in locating a participating retailer, e-mail B2B@penguinrandomhouse.com.

Library of Congress Cataloging-in-Publication Data

Names: Purdy, Chase, author.
Title: Billion dollar burger: inside big tech's race for the future of food / Chase Purdy.
Description: [New York] : Portfolio/Penguin, [2020] |
Includes bibliographical references and index.
Identifiers: LCCN 2019056850 (print) | LCCN 2019056851 (ebook) |
ISBN 9780525536949 (hardcover) | ISBN 9780525536956 (ebook)
Subjects: LCSH: Meat --Moral and ethical aspects. |
Meat industry and trade—Environmental aspects. | Cultured meat.
Classification: LCC GT2868.55 .P87 2020 (print) | LCC GT2868.55 (ebook) |
DDC 641.3/6—dc23
LC record available at https://lccn.loc.gov/2019056850
LC ebook record available at https://lccn.loc.gov/2019056851

Paperback ISBN: 9780593853863

BOOK DESIGN BY NICOLE LAROCHE

147141878

To my mother

CONTENTS

AUTHOR'S NOTE

Throughout this book I use the term "cell-cultured meat" or "cultured meat" because it is a scientifically accurate descriptor that reflects how the product is made. Even still, scientists, entrepreneurs, and regulators have not yet unified around a single term.

PROLOGUE

t's rare, but every now and again the biggest solution is incredibly, incredibly small.

"Everyone sitting here with bated breath is dying to see what's underneath," says Nina Hossain, a British television personality. It's August 2013. She's on a brightly lit stage with four others. She turns to Dr. Mark Post, on her left, who has flown to London from Central Europe. He's staring down at the big desk before him, at a silver platter covered by a silver dome. "So can you do the honors and lift the lid on your creation?"

It was a make-or-break moment, proof that the concept could be tugged from the imagination and into reality. And maybe someone rich would see it.

"It's, well, that's what we need, it's money," a scientist on Post's team would later say. "And I don't care who it is, if it's Bill Gates or Paul McCartney or whatever, but someone to really see, literally see, that there's a future behind this process."

Post reached down and lifted the lid. Cameras in the room zoomed tightly onto a little petri dish filled with pink ground beef—only this wasn't ordinary beef. It was grown in Post's laboratory, painstakingly cultured from microscopic cells that had been collected from a living cow—a cow that was still alive, not needing to be slaughtered for the sake of Post's burger. The five-ounce patty cost $330,000 to produce.

Chef Richard McGeown cooked the burger onstage. Sitting next to Post were Hanni Ruetzler, a food scientist, and Josh Schonwald, a food writer, who craned their necks to watch the burger sizzle in a searing-hot frying pan. McGeown served it up next to a lettuce leaf, a bun, and some tomato slices.

Ruetzler sliced into the burger. She pricked it with a fork, bringing it up close to her face to catch its aroma and inspect it closely. Then she took a bite, chewing slowly.

"I was expecting the texture to be more soft," she said. "There is quite some intense taste. It's close to meat, but it's not that juicy. The consistency is perfect, but I miss salt and pepper."

Then Schonwald did the same.

"The mouthfeel is like meat," he said. "I miss the fat, there's a leanness to it, but the general bite feels like a hamburger."

The BBC, *The New York Times*, *The Telegraph*, and National Public Radio dispatched stories. Twitter and Reddit flared up. In the face of a changing climate, growing concerns about animal welfare, antibiotic resistance, and global hunger, maybe, just maybe, we could take something small and use it to make big steps toward reshaping animal agriculture, a system that is responsible for about 14 percent of the globe's greenhouse gas emissions and the yearly slaughter of some 65 billion animals—not including fish.

Triumphant at the success of his showcase, Mark Post packed up and flew back home to the Netherlands. He returned to his Maastricht University laboratory where he got back to work, quietly ushering his expensive little burger into its next phase.

And then nothing. Silence.

A couple years ticked by and nobody heard much more about cell-cultured meat. Meanwhile, the general public began to catch wind of high-tech plant-based burgers made by the likes of Beyond Meat and Impossible Foods, imitation meat that put traditional garden-variety black bean patties to shame as they were marched into grocery stores and showed up on the menus of high-end restaurants like David Chang's Momofuku Nishi in New York's moneyed Chelsea neighborhood, and eventually fast-food chains as commonplace as Burger King.

But the science didn't stop. And slowly, cultured meat began to creep back into the public consciousness. Post founded his own company and called it Mosa Meat. In 2015, Uma Valeti followed suit and founded Bay Area–based Memphis Meats. Yuki Hanyu started Tokyo-based IntegriCulture. Ido Savir founded Israel-based SuperMeat. Then Finless Foods, JUST, and Aleph Farms announced their own laboratories a year later, followed by Mission Barns and Future Meat Technologies a year after that. Slowly, stories and photos have started trickling out from these laboratories, showing chicken tenders, bluefin tuna, burgers, foie gras, steaks, and nuggets. Investment in these companies is picking up, too. Since 2015, more than $100 million has been invested in cell-cultured meat by venture capitalists and existing food giants—with the number and size of deals increasing every year.

Buoyed by venture capital and billionaire investors looking to overturn the multibillion-dollar meat industry, these start-ups are in an edible space race to get cell-cultured meat to market first. It's a complicated problem. First they need to devise a way to scale cell-culture technology at a reasonable cost. In 2013, it was priced at $1.2 million per pound. Now it's hovering around $50 per pound, a precipitous drop as the technologists behind it have pushed the science to new heights. Then they need to hack a path forward through byzantine regulations designed around the needs and desires of the trillion-dollar meat industry. Against that kind of economic power, even the best-funded start-ups look like lightweights.

But one of these companies claims it'll beat everyone else to the finish line. Josh Tetrick is a vegan activist and the CEO of San Francisco–based JUST. He aims to reshuffle how humans around the world think about the meat they consume, where it comes from, and how it's made—and he has one big advantage over his competition. He's the only one who actually knows how to run a food company. Since 2011, he's successfully rolled out a line of vegan condiments, vegan cookie doughs, and a popular liquid egg alternative made from mung beans. Those have been adopted globally by stores and restaurants. He has already-established relationships up and down the food supply chain and the industry know-how to more easily get future food products onto the market where they'll go head-to-head with the conventional meats we've always known. And he's investing—heavily—in the science behind cell-cultured meat, building an on-site laboratory with a staff of about a dozen scientists who are experts in this new field.

The science is complex. The $1.8 trillion meat market is hypercompetitive. The stakes are high. If he messes up, it could scramble the rollout of one of the most promising answers to the climate crisis yet. But if he succeeds, it could begin a revolution.

BILLION
DOLLAR
BURGER

DIGGING IN

I slide my pen and notepad to the side. It's a sunny afternoon in 2018. Scooting my metal stool up to the edge of a tabletop in JUST's impressive food laboratory, I peer around me. This thing I'm waiting to see is supposed to be the powder keg that sets off a culinary shift—much talked about, rarely seen, and almost never tasted.

The company's headquarters sits in the heart of San Francisco's Mission District, at the corner of Sixteenth and Folsom streets. It's two stories tall and sprawls across 98,000 square feet of space—all enclosed behind thick, pinkish-beige walls that once housed the Joseph Schmidt Confections chocolate factory and later a Disney Pixar studio. JUST has married the building's food and technology pasts. Each day more than one hundred employees show up to work here. They enter through a front door of frosted glass, covered with the blemishes of bygone years and surrounded by a smudged golden frame. The entryway leads up a flight of black stairs, down a carpeted hallway, past a company break room—well stocked, of course, with avocados and La Croix flavored seltzers—and then finally into the company's cavernous nerve center. The floors are made of slick concrete, the walls are white, and dozens of long wooden desks stretch across the center of a room that's bathed in natural light pouring through a wall of tall windows.

All around me, teams of scientists are using expensive lab equipment to dive deep into the inner lives of plants, exploring the tiny components that make them unique in an effort to figure out how they can be used to make new types of food. What plant proteins can work as emulsifiers? Which ones can withstand stovetop heat? In a corner of the lab, a half dozen robotic arms sit in big clear boxes, a few of them swiveling to and fro, analyzing all manner of foodstuff; testing and cataloging the different aspects of plants that have been collected by JUST from all across the world in hopes they'll prove beneficial.

In the center of the room, not far from where I'm sitting, several sofas are clustered together, occupied by one of the many dogs that regularly lope throughout the company's headquarters on a given day. If anything is certain about JUST, it is that you can always count on seeing a dog.

I'd heard about this food before, of course, the stuff I've come to taste. It was only five years earlier that Mark Post had unveiled his burger. In those years between, pictures of early renditions of cell-cultured meat haven't been too tough to find, but even as a reporter charged with covering the global meat industry, I still hadn't seen any cell-cultured meat up close for myself, let alone had the opportunity to taste the stuff.

When it does arrive, it's on a small white plate, artfully wreathed by a fan of toasted sourdough and a distracting sprig of greenery. A lab technician slides the plate in front of me and I let it sit there for a moment.

"More people have gone into outer space than have tried clean meat," he tells me. More than five hundred people had been in space at that point. Only dozens could say they had tasted meat grown from animal cells. I was about to be inducted into an exclusive club.

I repeated the line to myself as I stared curiously at my plate.

"More people have gone into outer space. More people have gone into outer space . . . outer space."

I reached for a slice of toast, took a small butter knife, and dipped it into the main attraction: a dollop of golden-beige paste. Pâté had never appeared on any kitchen table during my childhood in Louisville, Kentucky, and I'd never acquired a taste for it.

Am I ready for this? I wondered, scraping the pâté across the bread. Until this moment, probably like you, every bit of meat I'd ever tasted had come from a once-sentient creature.

This particular dish was grown from duck cells. A lot of the companies growing cell-cultured meat chose to tackle meat from birds first. Avian cells often grow better in cultured settings than mammalian cells do. For starters, it's easier to tinker with them and get them to do what you want—they have better plasticity. Mammal cells are harder to coax into compliance; source animals must be young because they have healthier muscle stem cells, for instance. By contrast, scientists have learned that muscle stem cells, sometimes even from older birds, are still efficient for proliferation in a laboratory setting.

I bring the toast up to my mouth and take a bite, chewing for a moment, savoring—and judging—the pâté's silky texture and rich aftertaste. My eyes, mouth, and nose instinctively registered what I was eating as meat—but crackling across the buzzing synapses of my brain was only one thought: *Cells!*

One of the men in cream-colored lab coats, a professional chef who had been hired as a product engineer, searches my face for a reaction. His name is Thomas Bowman, he's been recognized by Michelin and, at the time, was leading product development in JUST's test kitchen. That's a long-winded way of saying that he took what the company's cell-cultured meat lab produced and tested it in the kitchen, providing a model for how real consumers might prepare and cook it.

I leaned forward for another helping, smearing more pâté onto a second slice of bread. As I chewed, a competitive game of tug-of-war between cold logic and cautious emotion unfolded in my mind. Of course, everything we eat is ultimately just a bunch of cells—the bananas we peel, the steaks we marinate, the potatoes we boil and simmer—and yet in that moment the thing I was swallowing seemed somehow alien. The pâté placed before me was the result of decades of scientific tinkering and the subject of science fiction; a dream product in the eyes of scientists and activists—foie gras grown not inside the tortured liver of a duck, but inside the walls of a Silicon Valley laboratory.

It was surreal but deeply compelling. As I would find out over the course of the following year, that pasty clump of cells spread over toast represented something much bigger and more globally significant: a preface to a growing food movement that's seeking to provide an ethical solution to the many unethical problems of the modern food system. By harvesting animal cells and quite literally growing them into fat and muscle tissue inside industrial bioreactors, humans have figured out how to create the exact same meats we've eaten for more than half a million years.

In doing so, those scientists hope to enable us to sidestep the need to slaughter billions of animals annually, and theoretically, in time, eliminate the need for an industrial farming system that pumps an alarming amount of greenhouse gases into the Earth's warming atmosphere each year. Scientists agree that animal agriculture is responsible for about 14 percent of greenhouse gas emissions.

Fully wrapping our heads around the impact of the animal agriculture system we've always known is mind-bogglingly difficult. Lots of scientists attempt to measure the full environmental footprint of animal agriculture, and almost all of them have run into fierce sets of critics who challenge their methodologies and motives. Did the scientist measure the life cycle of a single animal and then multiply those data to represent its specific sector? Did they include data on the energy used to grow, manage, and transport the feed grain for cows, pigs, chickens, and other animals? How about factoring in deforestation to make room for grazing? Or the long impact of water pollution from nitrous oxide in manure?

Those working on cell-cultured meat understand that by addressing these issues, they'll be helping to bring that number down; they'll be pushing back against the planet's march toward a full-blown climate crisis, one that already promises to reshape where and how humans live.

In 2006, the United Nations Food and Agriculture Organization published a landmark report that estimated animal farming is the source of some 18 percent of total greenhouse gas emissions, more than all the cars, ships, trains, and planes that crisscross the globe belching smoke. The report said 9 percent of the globe's human-related carbon dioxide emissions could be tied back to animal agriculture, 37 percent of methane emissions, and 65 percent of nitrous oxide (mostly from cow manure).

The report got a lot of attention and triggered a wave of people—activists and entrepreneurs alike—to take action. But it didn't go unchallenged, and since then scientists have produced more nuanced findings.

The beef and dairy industries get most of the attention, as they contribute 41 percent and 20 percent of the sector's overall greenhouse gas emissions, respectively. The pork industry creates about 9 percent of emissions, and the broiler chicken and egg industries collectively contribute about 8 percent. Cows do more damage because of their digestive process, which, through fermentation in their four stomachs, creates a lot of methane gas. Once released into the atmosphere, methane has about twenty times the heat-trapping power as carbon.

Just how much methane a single, 1,200-pound cow produces each year depends entirely on what the animal is fed, but scientists often cite about 100 kilograms of methane, about the same as a car burning through more than 230 gallons of gasoline.

Suddenly, the prospect of finding an alternative way to create real meat becomes even more enticing.

Then there's the inherent wastefulness of animal agriculture. It takes about 6 pounds of animal feed to produce 1 pound of beef, 3.5 pounds of feed for 1 pound of pork, and 2 pounds for a single pound of chicken. Animal agriculture relies on growing plant protein—vast fields of corn and soybeans—only to cycle it through an animal that has to be killed to yield less weight in food than the plants it ate. Of course, plant protein doesn't have the same nutritional profile as animal protein. But these numbers are the foundation of the compelling case for reducing consumption of meat in order to be better stewards of the planet, especially as estimates show the number of mouths to feed on earth—10 billion by 2050—is steadily rising.

Environmental sustainability is a compelling reason to consider giving up meat, but there are other reasons, too, especially animal welfare. The farms that raise chickens and pigs have gotten bigger and more opaque, not smaller and more transparent. Data collected by the US Department of Agriculture show the number of total farms in America has declined from 6.8 million in 1935 to fewer than 2.06 million in 2016, the result of a jaw-dropping amount of consolidation that has wiped out family farms. That's nearly a 70 percent decline. USDA data show the American dairy herd shrank by about 60 percent in the last ninety years, but only because humans figured out how to jury-rig individual cows to produce more than 400 percent more pounds of milk fat per head, according to data compiled by the USDA's National Agricultural Statistics Service. And the fishing industry remains one of the world's murkiest, with little over-sight. We have depleted vast stores of the planet's fish popula-tion (we kill an estimated 1 trillion fish a year); meanwhile, the fishing industry has relied on slave labor.

Bowman sets another plate before me, this time bearing a more familiar sight: tacos. In front of me are two corn tortillas topped with cilantro and postage stamp–size bits of duck cho-rizo. It was grown from the same batch of cells used to make the pâté. I tilt my head, peering at the plate at tabletop level and playing inspector, as if I might discover something amiss. I imag-ine throwing a finger into the air to say, "This is close, but the meat looks a bit rubbery."

But of course I don't. Any restaurant could have served the same dish to me and I would have been fooled. The seasoned cho-rizo tasted and felt just as I might expect, as moist and richly fla-vored as anything I'd get from the myriad taco stands outside JUST's office in the historically Latino Mission District. And on a biological level, there was no difference between my lab-concocted taco and the chorizo I could get at the taco stand outside.

Are we ready? I wondered.

Josh Tetrick is the kind of person who forces you to pause and squint. There's a quiet anxiousness inside him, making him at once curious and confusing. And it oozes into the work he's doing.

In December 2011, he cofounded his company with his child-hood best friend and fellow vegan activist, Josh Balk. At the time it was named Hampton Creek (it would later be rebranded under its aphoristic current name, JUST), and it had a simple goal: to make the egg industry obsolete by developing convincing plant-based liquid eggs.

It took longer than expected for Tetrick to finally get the recipe for that product right, but in the years between he has managed to make a name for himself and his company by creating successful lines of vegan condiments and cookie doughs that today can be found in some of the world's biggest grocery stores. In doing so, he grew a company that once operated out of a garage into Silicon Valley's first—and still one of its few—food technology unicorn—a start-up with a valuation of more than $1 billion.

Condiments, egg alternatives, and cookie dough helped cushion JUST with capital, but those products make only small nips at the heels of what Tetrick describes as an entrenched and degrading food economy built on animal suffering. So three tumultuous years after his company got off the ground, he announced that, not only was he joining the spirited global race to make lab-grown, cell-cultured meat a reality, but also he would be the first to get it to the market and in front of everyday people.

And this is what brought me to his California laboratory—to learn more about his work, his motivations, and what this new food might mean for the culture of cuisine.

Tetrick is often seated near the center of the main workroom. He's rarely at a desk, opting instead to sink into one of the room's several sofas. Elie, a loyal golden retriever named after the American-Romanian Jewish writer Elie Wiesel, is lying at his feet. And when Tetrick does buzz from one section of the laboratory to another, Elie traipses alongside him, ever loyal and obedient.

Tetrick's tuft of unkempt hair and sharp nose would give him a sparrow-like quality were it not for his imposing size. A muscled former athlete, he stands just under six feet tall and has managed to keep a figure vaguely reminiscent of his former days as a high school and college football player. Unlike many of his peers at rival companies, he has no scientific background at all. Nor does he have any education or prior experience in nutrition, business, or management. He is, in fact, a lawyer by training and an activist at heart—not exactly the type of person I expected to find at the helm of a company working to push food science to new heights while aggressively searching for ways to deal a punishing blow to the current meat industry. In many ways, he doesn't fit in this ecosystem. And even today, if you grab a drink with investors and other entrepreneurs in Silicon Valley's food technology scene, the name "Tetrick" often elicits eye rolling and groans.

He's a marketing maven with a skimpy science background who somehow found his way to the top of a relatively tiny field in the world of food. His success is undeniable, even to his harshest critics. And one thing is for sure: He talks a good talk.

"I think at its simplest, food sucks," Tetrick tells me. His words tumble out so rapidly and confidently that it's clear he's given this soliloquy before.

"It sucks because we harm animals, it sucks because we harm the environment, and we harm our bodies. So we're trying to figure out a way to make it not suck as much. That means taking animals out of the equation entirely."

Though Tetrick might sound radical, even industry insiders are coming to agree with him. At the United Nations General Assembly meetings on climate change in September 2019, Emmanuel Faber, the chairman and CEO of Danone, one of the world's largest food manufacturing companies, uttered something rarely heard from the mouths of food industry titans.

"The food system that we have built over the last century is a dead end for the future," he said.

To that end, it's encouraging that the landscape of food companies in recent years has given way to a legion of new start-ups, many of them grappling to tear at the seams of the status quo and capture the attention of consumers in the process.

The American market, in particular, is already familiar with a plant-based approach to replacing meat. Beyond Meat and Impossible Foods have certainly raced to the forefront of the meat alternative market, where they've also caught some heat. From a nutritional standpoint those products' healthfulness is questionable. A single Impossible Foods burger patty, for instance, contains about 16 percent of a person's recommended daily value of sodium. A conventional lean-beef patty contains about 1 percent.

The food I'm most interested in, the kind Tetrick is growing from cells, is far more intriguing than any plant-based imitator. The cell-cultured meat companies promise products with the same nutritional profile as meat that are simultaneously better for the planet. And cell-culturing food could wind up leading to so many new types of foods that we haven't even dreamed up yet, products that extend beyond meat. A firm called Clara Foods is creating egg whites that never needed a live chicken. Perfect Day has trained tiny specks of yeast to make the same casein and whey protein that was once available only from cow's milk. One company, the Berkeley-based Wild Earth, has even experimented with growing cell-cultured mouse meat for cats. And those are only a handful of start-ups in the growing food tech sector.

Some scientists in this space say that, like the agricultural and industrial revolutions before it, cellular agriculture is the next evolution in how we make food. And it's not so hard to imagine that it will lead to new types of products, much the same way that fermentation opened the door for us to make cheese, yogurt, and beer.

There are about nine such cell-cultured meat companies that have caught my interest, all of them leading the charge to make cell-cultured meat and fish a reality, each a player in a fast-paced edible space race to get this new food to market first. Four are in Silicon Valley, three are in Israel, one is in the Netherlands, and one is in Japan.

Together, with a mountain of money, the attention of influential billionaires, and help from some of the world's biggest food companies, these start-ups have made incredible strides in the last decade. They've developed the technology and have transformed it from a science fiction trope into an edible product that's fast approaching the point of being an economical reality. In October 2019, Tetrick said the cost of one of JUST's pure cell-cultured chicken nuggets had gotten down to $50 per nugget—or about $1,000 a pound. It's laughable at first glance, but consider the fact that just six years before cell-cultured meat was going for $1.2 million per pound and one can begin to understand just how quickly these companies have been able to decrease costs. And while other companies in this space compete with Tetrick on science and technology, he has an advantage that none of his peers share. Most of them are still just laboratories, while he's already the CEO of a fully operational food company with relationships built with retailers and supply chain players across the world. It arguably gives JUST a major edge once its cell-cultured meat is ready for consumers.

The concept is one that scientists have kicked around for decades. The people behind it often reference former British prime minister Winston Churchill when talking about their work. That's because in March 1932, Churchill wrote an article for *Popular Mechanics* magazine in which he offered the optimistic vision that, "fifty years hence, we shall escape the absurdity of growing a whole chicken in order to eat the breast or wing by growing these parts separately under a suitable medium."

Humanity missed the mark by a few decades, but Churchill's inkling has finally come to pass. And according to the venture capitalists, CEOs, and activists pushing this new food into fruition, cell-cultured meat products couldn't come a moment too soon.

The nascent food technology sector wants to be the answer to global problems. And so this small constellation of new food companies—including JUST—represents the high-dollar, low-key movement that's chugging into our lives with a sense of principled urgency. The companies toying with cell-cultured meat often cite speculative data presented in a 2011 Oxford University study when saying that their processes will require 45 percent less energy, produce 96 percent less greenhouse gas, and use 99 percent less land and 96 percent less water than the current agricultural operations producing our meat and dairy foods. The truth is that nobody yet knows the exact percentages for how much more environmentally friendly this technology will be, but few dispute that, generally, it will be a game changer for the planet.

Humans currently spend about $750 billion a year on meat

and about $330 billion on dairy products alone—the idea is that there's plenty of opportunity to slice into that market share. Financial analysts at Rabobank, a global leader in food and agriculture financing, say snapping up just 5 percent market share would signal a major disruption to the existing system.

But the monolithic global meat industry isn't particularly keen to step aside and make room for new market entrants. In the United States, the beef industry in particular has pushed back against Silicon Valley's high-minded technologists. The powerful cattlemen's lobby, along with those representing the pork and chicken industries, has exerted much influence in Washington, DC, to ensure that they play a role in shaping just how cell-cultured meat will be regulated, all while many of their members trash the product by calling it "fake meat."

Initially, it was the science that kept cell-cultured meat products from the market. Making cultured meat was just too expensive and the technology wasn't able to produce foods that were analogous to conventional products. But as the start-ups got better at their craft and the quality of their products improved, the biggest barrier to the market changed. It's a difficult regulatory landscape that's become the chief obstacle stopping people such as Tetrick from getting cell-cultured meat into markets around the globe and in front of consumers.

I look down at my plate, which is now empty save for a lonely pair of cilantro leaflets. As I push my stool away from the lab table at which I've been eating, I've got one thing on my mind: It's autumn 2018, and a man named Josh, an Alabama-born, Philadelphia-raised vegan activist with scruffy hair and a razor-wire stare, believes he is poised to change the culture of food as we've always known it. He doesn't wear a suit and he isn't interested in playing nice. He's been strategizing a plan that's about to walk him onto a long-haul flight halfway around the world to carry out what has been, until now, an impossible feat. He's up against the herculean forces of an entrenched industry that isn't eager to share the market, and the habitual culinary traditions of civilization that have been developed across thousands of years. And there, at the forefront of my brain, is a single jagged question, perhaps the most human question to have at this moment in time.

Is he ready?

2

A CULINARY
GODFATHER

n his 1984 novel *The Neuromancer*, William Gibson lays out a
scene in which one of his futuristic cyborgs, Molly, unceremo-
niously snatches a cut of conventionally grown steak off someone else's plate.

"Gimme that," she says, grabbing the dish. "You know what
this costs? They gotta raise a whole animal for years and then
they kill it. This isn't vat stuff."

In Margaret Atwood's 2003 work *Oryx and Crake*, she introduced readers to a truly macabre future food, ChickieNob: a live
chicken engineered without any sort of typical head or even a
beak; a freakish avian creation designed to grow up to twenty
breasts at the same time.

Only time will reveal whether Gibson's or Atwood's fictional vi-
sions will be far off the mark of what cell-cultured meat companies

will offer us in real life. "Vat stuff" isn't so different from what to-day's earnest young collection of start-ups are hoping to usher into reality.

It is inviting to think of cell-cultured meat as the stuff of science fiction, brought to life by an ingenious collective of hoodie-donning technology bros operating out of tiny Silicon Valley coworking spaces. In truth, though, the scientific effort to bring cultured meat to market has been sputtering along for several decades, well before Atwood and Gibson ever thought to put their ideas down on paper.

At the heart of this movement is a solemn figure named Willem van Eelen. Though he passed away in 2015 the Dutch entrepreneur and doctor is still regarded by scientists working in the field as the godfather of cell-cultured meat, the person who prodded and persisted—at times bitterly—within the scientific community to try to convince people to take the idea seriously. His doggedness came from a deeply personal place, chiseled into his worldview during moments of despair, when he was very young, and many thousands of miles away from his home.

Van Eelen was relentlessly practical. The central purpose of cell-cultured meat was to solve the problem of hunger. In some respects, it makes sense the concept is covered in Dutch finger-prints. After all, it was the Dutch who had the spirit and ingenuity several hundred years ago to defy the natural world, erecting a system of dikes and canals to keep floodwaters at bay. The Dutch have a history of pushing back against nature, and much of van Eelen's life was spent doing the same.

Willem van Eelen was born on July 4, 1923, in colonial Indonesia, to Gerard and Caroline van Eelen, an eccentric pair of Dutch expatriates. Caroline was a gifted pianist who would play occasional concerts, and Gerard was a Lutheran missionary who initially worked in a remote Indonesian city where he focused his attention on constructing a church and converting the local population. He would go on to independently establish and run a recognized leprosy colony in Indonesia, and eventually would become a high-ranking official in the Red Cross. Van Eelen grew up the youngest of four children—he had one sister and two brothers—and, according to his daughter, Ira van Eelen, forever worked to live up to the standards set by his father.

I visited Ira at her home in Amsterdam, where she lives with her husband and teenage son on a large, personally refurbished and designed houseboat that's docked in one of the city's picturesque east-end canals. The sun pours through large skylights and numerous portholes, revealing a warm and tidy combined kitchen and living room. Ira sits at the end of a marble-top kitchen island, beneath two hanging lights that sway ever so gently with the rocking of the boat.

"This is the cell-cultured meat origin," she says, sliding across the tabletop a black-and-white photo showing four stern figures— her boyish father, his two brothers, and her grandfather. Ira pushes a few strands of her blond hair behind one of her ears and peers down through wire-framed glasses. Running her finger from left to right, she begins her family's story.

One of her father's older brothers would go on to work as a

reclusive doctor in the far reaches of northern Norway. The other became a male model and grower of bonsai trees. Both of them bullied van Eelen relentlessly in their youth, imbuing him even more with a desire to impress. Missing from the photo was van Eelen's mother, Caroline, and sister, Maria, whom Ira described fondly as a fellow eccentric who "had affairs, rode horses, went all over the world" and would ultimately pick up and leave her humble life in the Netherlands, decamping to India where she joined the enigmatic and mystic guru Bhagwan Shree Rajneesh, who was also more popularly known as Osho.*

When he was in his early forties, Gerard van Eelen moved his young family out of Indonesia to eastern Holland for a brief period as he studied in Leiden, the birthplace of the tulip, to obtain his medical degree. Once finished, he moved them back to Indonesia, only this time to the posh city of Bandung in West Java, where he worked as a doctor. Their lifestyle became posher as well: regular tennis outings (they had their own court), winter skiing trips in Davos, Switzerland, and shipping the kids back and forth to the Netherlands' elite preparatory schools.

Their charmed life was disrupted by the creeping threat of Adolf Hitler's Nazi Germany. Van Eelen was fifteen and at school when he received an urgent telegram from his parents in Indonesia, beckoning him back to Asia posthaste. It was an attempt to prevent him from getting ensnared by the oncoming danger that would swiftly and unsparingly ravage Europe. Van Eelen

---

* **Osho:** Subject of the Netflix documentary *Wild Wild Country.*

was able to find one of the last ships en route to Indonesia, where his family believed he would be safe. On May 10, 1940, German troops invaded Holland.

Van Eelen's father and brothers were involved in the war effort, members of the Red Cross and the Dutch Royal Indonesian Army. To impress his father and then-girlfriend, van Eelen decided to enlist himself.

His safety outside of Europe was short-lived, though, as the Japanese military began its takeover of Indonesia.

The war left two scars on van Eelen. The first was a cautiousness toward films that featured airplanes flying toward the screen. During his time in the Dutch military, he served on a mountaintop that overlooked the sea. He and his fellow soldiers were assigned to stand guard over and operate a massive cannon, trying to blow Japanese fighter planes out of the sky as they buzzed, one by one, like predatory wasps down toward the Dutch settlements. He later recalled to his family the horrors of being rained upon by machine-gun fire, thousands of bullets ricocheting off the concrete surfaces surrounding the cannon pit, where comrades were often caught in the strafing and perished.

The second scar was a complicated relationship with food. Five days after the Germans marched into Holland, the Dutch government surrendered. Less than two years after that, in March 1942, the Dutch officially capitulated to the Japanese, forcing van Eelen and his fellow troops in Indonesia on a deep retreat into the surrounding jungle. For many, including van Eelen, the humid jungles proved about as perilous as facing the

Japanese head-on. Van Eelen surrendered as a prisoner of war to the Japanese, beginning a tortuous three-year stint that he later nicknamed "Travel Agency Nippon." The Japanese military would frequently move their prisoners by the thousands, transporting them in cruel and inhumane conditions from island to island to clear land and build airstrips.

In the stories he would later tell his children, van Eelen recalled the devastating effects of the starvation that unfolded before him during his internment. According to his recollections, the prisoners divided into two groups: those who lost their grip, and those who held on. As desperation and extreme hunger set in, van Eelen watched as the prisoners around him gave in to their most basic instincts, tearing apart any animal they could find.

The experience affronted his sense of human dignity, and it attuned him to the urgency of starvation. It was a problem to solve. So in 1948, after the war had ended, van Eelen returned to the Netherlands and enrolled in medical school.

While touring the laboratories one afternoon, he walked by a room in which a group of scientists were tinkering with a piece of tissue taken from an animal. They were attempting to keep the tissue alive, to get it to grow. Success would open the possibility of one day growing organs for humans. He and his fellow students looked on with fascination, but the wheels turning in van Eelen's head had nothing to do with human health. He was asking an entirely different question: Could we grow food the way we grow tissue?

It wasn't that outlandish an idea. The medical field has grown cells in laboratories for all sorts of reasons. Cell culturing is a foundation of modern biology and medicine, a tool that has contributed to our understanding of physiology and our drugs to combat AIDS, cancer, and much more. As far as van Eelen was concerned, if you could grow an organ in a petri dish, why couldn't you grow muscle and fat? And if you could grow flesh, why couldn't you eat it?

"The only thing I saw was a piece of meat," he would tell his daughter about the moment the lightbulb flicked on.

Van Eelen may have thought up the concept of cell-cultured meat, but he didn't have the expertise to translate the idea into reality. He spent the latter half of his life—much of it through brooding fits of impatience—trying to corral the scientific talent and funding that he needed to bring his dream to fruition.

Van Eelen had presence. He could fill a doorway. And he wasn't bashful about showing up—sometimes uninvited—at the offices of wealthy entrepreneurs to describe his vision of lab-grown meat. It was a hard idea to pitch, as the underlying research was almost exclusively considered a medical technology in the Netherlands. Though extremely slow going, his commitment to the calling did wind up getting some traction. By the early 1990s, van Eelen had raised close to $750,000 from a handful of investors.

In 1997, he filed for the first of several patents relating to the process of culturing meat. It was approved in 1999. A few years later, in 2002, he formed a consortium of scientists that partnered with Meester Stegeman BV, then a division of the food company Sara Lee. Among this collection of scientists was Mark Post, whose impressive work would go on to cement him as a major player in the cell-cultured meat movement.

Around this time, van Eelen ran across the name of a man in America who had also been filing for patents on similar technology. Curious, he tracked down the man's telephone number, picked up a phone, and waited impatiently.

In the middle of the night, on another continent, the phone at the other end of that call began to ring.

Stirred from sleep and groggy, a man in Los Angeles answered the buzzing receiver. He looked at the clock, noticing it was some time well after midnight. He recalled hearing what on the other line sounded at first like a thick German accent.

"Is this John Wayne?" van Eelen barked.

"Who is this?" the man replied, perplexed.

"My name is Willem van Eelen, and we'll either be the best of friends or the worst of enemies."

He'd found Jon Vein's name attached to recently filed patents for lab-grown meat technology. It seemed he'd discovered another enthusiast. Vein was the former chief operations officer of Artist Management Group and a future member of Hillary Clinton's 2016 National Finance Committee during her US presidential campaign. But he also had a passion for future food.

A self-described geek at heart, Vein has a double engineering degree and a law degree. In the 1980s he started reading an

24

array of articles that seemed at the time to be mostly unrelated: the state of world hunger, the environmental footprint of the animal agriculture industry, stories about avian influenza, deforestation for the sake of creating more grazing ground for cattle, and stories about tissue engineering.

He had the foresight to put all those pieces together, though, creating an obsession with culturing meat. If you could grow meat in a controlled setting, it might be a way to work around all the negative aspects of animal farming.

"So I brought on a couple of, at the time, well-known tissue engineers and asked if they could help me put my idea into writing," he said. And that's around the time he got his first call from van Eelen.

The two men became friends, and van Eelen grew to trust Vein. He believed Vein could help him make cell-cultured meat a reality, and even eventually sold Vein his patents to consolidate their intellectual property. Together, the two approached financiers and researchers in an attempt to get more people in academia involved in the mission.

Other work was happening in this field at the time, though it was primarily on the academic front. Van Eelen generally considered this progress too slow moving. In 2002, a scientist at Touro College in New York City named Morris Aaron Benjaminson managed to successfully culture the muscle tissue of a goldfish in a petri dish. The work was funded, in part, by a $62,000 grant from NASA to explore the concept of alternative food sources to help sustain astronaut crews while on long space journeys from Earth to the moon. Benjaminson cooked the goldfish muscle in olive oil, flavored it with spices, then presented it to a panel of people who came to the conclusion that the meat was, in fact, edible. You can imagine that, if ever it did materialize into an actual product, it could be a freeze-dried space meal that even the most adventurous astronaut might put off until well into their return journey back to Earth, when they're just too bored to continue pushing it to the back of the shuttle's kitchen cupboard. NASA never followed up on Benjaminson's study.

By then it was the mid-2000s. Interest in investing in cell-cultured meat was tepid. As van Eelen grew older and his health diminished, it became clearer that he might not survive to see the fruit of his labors, something that often cast him into gloomy moods.

Years later, on May 6, 2012, the president and founder of PETA, Ingrid Newkirk, wrote a small tribute to van Eelen in *The New York Times Magazine*, right on the cusp of the historic moment he'd been waiting for.

"Soon I will be able to eat meat again without any qualms, without worrying about my health, cruelty to animals or environmental degradation," she wrote. "That's because this autumn, 60 years after it was just a gleam in the eye of the Dutch entrepreneur Willem van Eelen, the very first laboratory-grown hamburger is to make its debut."

A few months later, Maastricht University professor Mark

Post—one of those early partners van Eelen had successfully enlisted into his consortium—unveiled his very early prototype of a lab-grown hamburger in London. It was comprised almost entirely of muscle tissue, and the cost to produce the burgers (only one was presented publicly but he made four total) exceeded $330,000, but it was progress.

Van Eelen, at the time in his late eighties, wasn't at the London conference. He read about the event the next day in the newspaper, unhappy at how much the burger reportedly cost. He didn't think it needed to be that expensive, his daughter said. He got a call from a journalist later that day—and it wasn't a pleasant one. According to Ira, her father went berserk. He had run out of money for more research, he wasn't convinced that Post's specific research was the right path forward, and the time he had left to create change was running short.

Despite his frustration, the fact that he could even read about cell-cultured meat being presented on a stage in London was a success story for van Eelen, who'd spent decades urging the world to take notice of a technology that he believed would someday be able to help eradicate hunger in a cost-effective way that was also environmentally sustainable.

Willem van Eelen was ninety-one years old when he died in 2015. At that time, none of today's cell-cultured meat start-ups existed. Now there are more than thirty, each pushing the science and economics of meat production forward. His shadow covers the entire present-day industry. Post, of course, worked alongside him. Memphis Meats CEO Uma Valeti, and others, have read much of his writing.

And Tetrick has acquired van Eelen's patents, a signal that the original dream is still moving forward, like a baton from one generation to the next, in a push to recast the future of food.

3

THE MOLECULAR
MIRACLE

This optimistic band of new food technology start-ups has made so much progress that, in just three years, the cost of producing some types of cell-cultured meat has drastically fallen. It took three lab technicians about three months to grow the twenty thousand muscle fibers in Post's $330,000 burger. A pound of that beef would have cost about $1.2 million if sold in a conventional grocery store. By March 2017, though, one of the start-ups, Memphis Meats, told *The Wall Street Journal* that it had gotten the price of cell-cultured chicken down to $9,000 per pound—still a humongous expense, but an exponential drop in price over the course of just a few years. In September 2018, at a food technology conference in Berkeley, California, the CEO of Memphis Meats announced the price had dropped again, this time to below $1,000 per pound. In early 2019, the Israel-based company Aleph Farms told a news crew that it had gotten its beef down to around $100 per pound. In October 2019, Tetrick said his pure cell-cultured chicken nuggets cost about $50 per nugget. The same month, Israel-based Future Meat Technologies said it was on track to have cell-cultured meat on the market in 2022 for about $10 per pound.

What makes cell-cultured meat such a costly venture? The brainpower it takes to push it into fruition, and its ingredients. At its simplest level, the recipe requires three things: cells, of course; a nutrient-dense liquid medium to feed the cells; and a sterile bioreactor that provides the right conditions to grow them.

The concept of culturing cells is pretty straightforward. By introducing a single cell to the right set of circumstances, it will naturally divide and duplicate many, many times. Once those cells multiply enough times, they organize themselves into a visible mass called tissue. Most of the meat we eat is primarily muscle tissue, which is about 75 percent water, 20 percent protein, 5 percent fat, with a trace amount of carbohydrates.

A whole mysterious world is swirling around within the cells themselves, each programmed by twisted strands of genetic code that can direct a single cell to respond to its environment, manufacture proteins, create energy, build skin, bone, and muscle, and make antibodies to protect itself from outside invaders, among many other things.

So how do the scientists at JUST and other companies get these cells? It's actually pretty simple. You can collect them by performing a harmless biopsy on a live animal. You can get them by harvesting tissue from the ovaries of a newly slaughtered animal (a good place to find young and strong cells). Some scientists have said you can even get them from the packages of meat found at any common grocery store, though that isn't commonly done and scientists debate how viable those cells would even be.

"The healthiest cells are going to be from a live biopsy and the healthiest of the healthiest cells are going to be from a biopsy of a very, very young animal," says Mike Selden, who leads Finless Foods. "Honestly, the grocery store stuff is totally useless because everything there is dead."

The scientists at cell-cultured meat companies identify which cells are stem cells. Whereas ordinary cells have limited utility, stem cells can divide and multiply many times, and they can transform into any of the more than two hundred types of cells that operate within animal bodies.

Think of our cells as individuals on a building construction site. Some are assigned to lay cement, some are trained as carpenters, and another might be an electrician. Together, they work harmoniously to build different components of the larger structure. But it is possible to pluck a cement worker from her job, send her to classes where she'll learn about wiring, plugs, and sockets. Then she can be sent back into the field as a newly trained electrician. The same thing can happen on a molecular level.

For instance, the scientists at JUST say that they can manipulate certain cells they collect from the pointy tip of a chicken's feather to divide and multiply not just into feather tissue, but also potentially into muscle or fat tissue. To do so, they subject the cells to a process called dedifferentiation, in which scientists can take a cell that would normally live its life as a feather producer, tell it to stop working on feathers, and then revert it back into an earlier-stage cell. It's kind of like sending a cell into a time machine, pushing it backward through its development and into what's called a pluripotent state. At this point, our one-time feather cell can be directed into choosing a new life adventure. Convincing it to take on a new, more specialized job—as a muscle or fat cell, for example—is called differentiation, the process by which a cell acquires the traits and tools it needs to perform a new set of functions.

Across the laboratories at universities and the cell-cultured meat companies, there are two different schools of thought on how to ensure that scientists have a steady supply of cells. The first is pretty obvious. Once every few months you take a trip to an animal farm to collect cells. Once you've got them, you make them multiply for as long as they are willing, and then once they stop you go out and collect more. Rinse and repeat. The second option is to take a group of cells and create what's known as an immortal cell line, which would eliminate the need to go out and perform a biopsy on a live animal.

That term, immortal cell line, is a bit deceptive. Its roots trace back to the 1950s, when most cellular biologists operated under the theory that, as long as you feed cells what they want and keep them in a condition that they like, they'll multiply forever. That's a tall order on its own for scientists, who have had to figure out just how to create those conditions.

But an American scientist named Leonard Hayflick upended that theory in 1962, when he showed that, inside every cell, there exists a sort of biological clock. And every time a single cell divides to create a whole new cell, that little clock counts each division. When a cell divides about fifty times the clock runs out. It suddenly decides to stop.

Also inside cells are threadlike chromosomes that carry genetic information. At the ends of these chromosomes are distinctive structures called telomeres. Like an important part of a machine, telomeres make sure that no crucial element of a cell's DNA is lost during the process of cell division. In essence, the telomeres hand out little instruction manuals so the cells know how to replicate correctly. The issue is that telomeres don't have an endless supply of instruction manuals, called bases. When a cell replicates over and over again, the loss of bases causes the telomeres to become shorter. Once its supply of bases is too low, the telomeres risk losing part of the DNA itself. The cell recognizes this threat as an error, and so the telomeres stop handing out instructions, which causes the cell to stop replicating.

This is a disadvantage for the makers of cultured meat, because it means a group of cells' ability to reproduce exponentially eventually tapers off. Since the 1960s, scientists, including those at cultured meat companies, have figured out ways to get cells to reproduce an even greater number of times. They commonly do so by supplementing cells' liquid medium with proteins and antioxidants that can lengthen their life spans. Now some companies are working with certain animal species' cell lines in which the individual cells can multiply well beyond fifty times.

Most of the cultured meat companies have chosen to pursue this route. Despite the difficulty of establishing immortal cell lines, doing so creates major advantages over repeatedly going back to an animal to harvest cells. It's easier to scale up the amount of meat that's grown because the cells don't peter out after fifty or so replications. Costs are lower because a company isn't using its resources, equipment, and manpower to collect the cells. And maintaining consistent meat quality is easier because the genetics of animals are inherently variable—not all cell populations will perform equally, and this can have a big impact on taste.

Creating these cell lines isn't easy, but thanks to collegiality within the scientific community, it's possible for some start-ups to avoid starting completely from scratch. They just need to find a way to obtain the beginning components of an established cell line. For companies in Silicon Valley (and the rest of the world), such a place exists on the other side of the country.

Outside Washington, DC, in Manassas, Virginia—nearly three thousand miles away from San Francisco—a nondescript building sits tucked behind Northern Virginia Community College, shrouded in trees. There's nothing notable about its exterior, which is angular, gray, and surrounded by small flower beds. But on the inside it holds a vast and priceless inventory of scientific wonders that will shape the future of food.

This is the American Type Culture Collection (ATCC), a nonprofit storage facility for all sorts of cellular cultures. Founded in 1921, the ATCC is a library of sorts, housing the world's largest and most diverse collection of human and animal cell lines. When researchers on the cutting edge of cellular biology create notable new cell cultures, it's a good bet that a sample of those will eventually make it to the ATCC's biorepository department for cryopreservation.

That makes the facility a veritable gold mine for entrepreneurs and scientists pushing the frontier of lab-made meats. Once a culture is added to the library's catalog—a temperature-controlled vault—individuals, researchers, and companies can pay for access.

Getting a culture into the ATCC library is a dream for Dr. Paul Mozdziak, who has been toiling for years in his laboratory at North Carolina State University to create what he calls an "infinite chicken cell culture." Mozdziak says his goal is to better understand the cellular and molecular mechanisms that control the growth of muscle tissue.

Mozdziak hopes his culture will be able to grow on its own in a bioreactor—a vessel for cell culture that looks sort of like a beer vat—where it literally floats, suspended in a liquid solution, constantly reproducing a viable mass of chicken meat without ever needing a bucket of grain feed, a barn, or a live chicken. And once that process is nailed down, the trick for scientists such as Mozdziak will be to move it outside a laboratory setting where it will be able to be produced in large quantities, scaled up to the point at which lots and lots of people can be fed.

On a theoretical level, there's nothing complicated about Mozdziak's goal of creating a scalable process. It's a matter of moving
from one-liter beakers to 100,000-liter bioreactor tanks. It's a
game of ratios.

There are dozens of future food companies playing with interesting new concepts for replacing animal meat products, but
they all run into the same problem: It's hard to affordably create
a lot of a product in a short amount of time because of the cost
of ingredients, but doing so is a crucial hurdle to overcome for
any company that wants to put a dent in the overall market.
Mozdziak hopes his research will help solve that problem for
people toying with lab-made chicken.

But before anyone can consider large-scale disruption, they
first have to figure out how to get those tiny little cells to multiply many, many times over. In order to do that at all, these building blocks of life need to be fed. This happens naturally inside
livestock animals (and in our own bodies), thanks to a vast and
intricate system of blood vessels. Blood delivers nutrients to
areas of our bodies. But when scientists grow cells in vitro, or
outside the body, they don't have that system of vessels. So they
improvise by immersing cells in a swimming pool of nutrient-
dense medium, the second of the three main ingredients needed
to grow meat from cells.

In the context of cell-cultured meat companies, the medium is
perhaps the most mysterious part of the process. It's a witch's
brew of macromolecular "growth factors" such as amino acids,
sugars, lipids, and hormones that cells need to survive and prolif-

erate. Also inside the medium—sometimes called serum—are hundreds of proteins that each carry out a specific task to help with the proliferation of cells. Among their many jobs is to transport insulin, which is necessary for growing many cultured vertebrate cells. Another is transferrin, which is a protein that delivers iron to cells in a culture. The most advanced start-ups have their own custom-tailored potions for the cells they use to produce their meat products; the recipes are closely guarded intellectual property.

For years, though, scientists in the cell-cultured meat space relied on fetal bovine serum as a medium. That's exactly what it sounds like: blood that has been drawn from the fetus of a cow. But the use of this serum was never going to work in the long run for this new crop of food tech companies. Four cups of the stuff can go for $1,150, and while it's great for growing relatively small quantities of meat in a laboratory setting (JUST scientists say you could grow about 10 kilograms of meat with that amount), it's not enough for the industrial-size vats the companies envision for producing enough meat to feed an entire market. At that price point, there's simply no way to affordably supply enough cow fetus blood to feed the meat cells that will be necessary to upend the larger factory farming system. And then, of course, there are the ethical hang-ups.

To eliminate the need for an animal agriculture system, the companies need to find an alternative to a material sourced from actual cows. To avoid using an animal-based product for the liquid medium, scientists at several cell-cultured meat companies have turned, of course, to plants. At JUST, Josh Tetrick has an edge over other start-ups in the space because of a singular database of knowledge his team of scientists has spent years creating. Its initial purpose was to create vegan mayonnaise and eggs, but now its applications for cell-cultured meat are much more significant.

The system is called Orchard, and it contains a bevy of valuable information about plant species from around the world—each one gathered, analyzed, and entered into the system as part of a project overseen by Udi Lazimy, who is in charge of raw plant sourcing and sustainability at JUST. Lazimy travels the world—visiting communities in places such as West Africa and India—collecting plants and bringing them back to San Francisco. Armed with data about protein richness among plant families and species, he focuses his efforts on areas known to cultivate diverse varieties of species that are known to have high protein content. He's collected more than two thousand plants from more than sixty countries, spanning every continent on the planet except Antarctica. When he and his team enter a new plant into the Orchard system, they note its characteristics: where it originated, how it responds to drought conditions, its allergenicity, its known flavors and odors, even the size of annual global production.

Much of the work at JUST is designed around figuring out what kinds of proteins are in different plant species. Most of a plant's protein can be found in its seeds, so Lazimy takes the seeds he collects and mills them on site into a fine powder, which is then stored in hundreds of meticulously labeled pet food containers in a temperature-controlled room on the first floor of JUST's headquarters.

Then, upstairs, scientists run samples of those powders through a series of high-tech machines—some worth hundreds of thousands of dollars—which analyze them and spit out data on all sorts of properties, including protein patterns and solubility. The company's scientists also test each powder for use in food— how it might impact the way something gels, or whether it would serve as a good emulsifier. In its earliest days, JUST's greatest ambition was to create a plant-based egg substitute. In 2018, it finally accomplished that goal when it released a liquid egg alternative based on the proteins found inside mung beans.

The company applied the same approach to figuring out how to make a liquid medium to replace fetal bovine serum. Using its database of plant proteins, JUST's chemists have engineered suitable liquid mediums.

But the plant-based mediums aren't perfect. The scientists at JUST may identify the necessary components for growing cells as they exist in fetal bovine serum, but the perfect potion that replicates natural mediums is elusive. The fact is that cells often don't grow as quickly in plant-based mediums as they do in fetal bovine serum. And even a near-perfect plant-based medium for growing duck cells may not be the right medium for cells from chickens or pigs. For that reason, the people in the company laboratory are analyzing and testing as many plants as possible, dreaming up new growth mediums and working to figure out how cells react once submerged in them. And though the new plant-based and synthetic liquid mediums are cheaper than fetal bovine serum, they are still expensive when considering how they might perform in large-scale manufacturing. Eitan Fischer, who worked as a scientist at JUST before starting his own company called Mission Barns, cautioned that, across the entire industry, the medium is the biggest factor in cost and the biggest hindrance to creating cost-competitive cultured meat. By the end of 2019, JUST scientists, led by Vítor Espírito Santo, said their in-house liquid medium was priced drastically below what they used to pay for fetal bovine serum, about $1 to $5 per liter, a $25 drop from the prior year and a far cry from the $1,200 that a liter of fetal bovine serum costs.

All that cellular activity takes place in the third and final item needed to grow meat: the bioreactor. Think of it as a mechanical cow—big containers in which scientists can create the conditions necessary for cells to grow once they are swimming in the liquid medium. These machines are usually cylindrical and can range in size from small flasks—which are commonly found in the start-ups' laboratories—to industrial-size vats. The ones used by Future Meat Technologies in Israel are about 600 liters—about the size of a refrigerator—and can theoretically grow the equivalent of 1,500 chickens in just a few weeks.

Once filled with the liquid and cells, the bioreactors are able to maintain a constant temperature, regulate pH levels, ensure that enough oxygen is moving around the tank, and manage nutrient concentrations and acidity. All of this is crucial to keep the cells blithely munching and growing in their medium.

There are five different models of bioreactors that cell-cultured meat scientists are using right now to grow their meat, each one designed primarily for medical research. But cell-cultured meat companies have different sets of needs from scientists working in medicine. Their work requires new kinds of designs, small engineering rejiggerings that don't exist just yet on the commercial market. The main challenge is creating environments inside the bioreactors that allow the cells to replicate with minimal disturbances, and that's tougher than it sounds.

As Jess Krieger, the founder of Bay Area–based cell-cultured meat company Artemys Foods, told me, cells like to grow together in clumps, thanks to the many types of connections that naturally occur between them. One is what scientists call a "gap junction," a special intercellular fusing that helps cells communicate with each other. Another is created by adhesive-like protein complexes called desmosomes. These connections help the cells grow together in a mass. In the bioreactors, these clusters not only cling to each other, but also find additional support by clinging to the walls of the tank or onto special platforms built into the bottoms of some bioreactor models.

The whole operation is tricky because scientists need to make sure that the nutrients in the liquid medium—as well as oxygen— are evenly distributed. To do so, the bioreactor creates a jet stream forceful enough to constantly stir the liquid, but gentle enough to do so without disturbing the cells. As can be imagined, the bigger the bioreactor, the more liquid is involved; and the more liquid that's involved, the stronger the force necessary to mix the medium.

Inside JUST, the company's chief technology officer Peter Licari led me into a small section of the cell-cultured meat laboratory, a space encased by clear plastic walls, where scientists Kevin Hicok and Vítor Espírito Santo are working busily with plates of cells from different animal species, monitoring their growth and sliding samples of them beneath powerful microscopes so that I might get a better look at how they form.

I peered through one and saw small pellet-shaped organisms; translucent little bubble creatures lining up alongside one another to create a larger mass. I inspected just a tiny sample, but once they've proliferated inside a bioreactor these cells can be harvested and processed into meat products that would be recognizable—think hot dogs, chicken nuggets, sausages, and burgers. When the cells are ready to be harvested, the scientists separate the cells from their medium by centrifuging them: sending the bioreactors whirling at high speeds. The liquid medium drains, spattering the solids against the walls of the reactor. Scientists then scrape out a dull pinkish-gray mass of cells with the consistency of smooth, wet Silly Putty. They need to be washed and centrifuged yet again before the cells are suitable for consumption, but it's a pretty quick process. Within several days the scientists can grow enough of these cells to have a handful of meat to package and send over to the product development team. That's when someone like Thomas Bowman, one of JUST's Michelin-star chefs, can begin working his magic.

To be sure, it is not necessary to have a Michelin-star chef on hand to make cell-cultured meat edible.

"What would happen if you ate it just like that, out of the centrifuge?" I ask.

"Nothing," Licari says, further describing how carefully calculated and monitored the process is from start to finish. "You could certainly eat that. It would be clean."

Clean. What a term.

For years I would avoid ordering burgers or steak cooked anything less than medium, an abundance of caution instilled in me by my mother, who didn't want to expose her children to the sundry microscopic visitors that lurk within raw meats.

Food-borne pathogens are notoriously difficult to pin down. Norovirus, salmonella, and *Escherichia coli* can find a way into a person through almost any food that hasn't been fully cooked or processed. They can travel as easily through a paper bag of supermarket flour as in a cheeseburger. Now, the scientists at cell-cultured meat companies say that, because they are creating their meat in totally sterile environments, their products are inherently safer and cleaner than the meats that come from the outside world, where conventional meat's exposure to the elements and animal manure has forced us to cook our foods at specific temperatures for safety reasons, burning off surviving microbes. If you really wanted to, you could theoretically eat cultured meat right out of the vat.

This idea that cell-cultured meat will be supersafe opens the opportunities for more than just new-age chicken, beef, and pork. In fact, if you talk to Isha Datar, the executive director of cellular agriculture nonprofit New Harvest, there are a lot of reasons to be incredibly optimistic about the benefits of culturing food. Those include potentially unlocking the ability to create new types of edible creations—food that goes beyond meat, even. And those aspirations are buoyed by the belief of scientists that cell-grown food will be safer to prepare and eat, in general.

"Everyone asks when lab-grown meat is going to be on store shelves, and to me that implies that there is an end point to the pursuit of cellular agriculture," she said. "But the end point isn't a product. This is much bigger: Cellular agriculture is a whole new industry—a whole new field of science, and a whole set of products that can change the way that food lands on our tables."

Datar is a vegan activist who runs New Harvest, an organization that, for years, has been pouring hundreds of thousands of dollars into the race to fund cellular agriculture research. She wants to get highly advanced, affordable, animal-free food products in front of consumers as quickly as possible—and she's focusing her organization's resources to get that done.

It's a bold vision for the future in which meat is only a starting point. Beyond it are the sundry new foods we have never even dreamed about yet. Certainly, foods that Willem van Eelen had never dreamed of.

The progress this tiny industry has made since van Eelen first started pushing it forward is astounding, but there's still so much science to perfect, so much ground left to cover and data to gather. And this goes beyond the details of what happens inside the bioreactors.

The cell-cultured meat companies make big claims about the potential benefits to the environment their technology could create. It doesn't require a huge leap to understand how the companies will use less land, water, and energy—but the fact is we still need to see more hard data before we can truly know just how much better for the earth this food technology will be. Academic analyses of cultured meat conducted thus far have been strictly theoretical. That's because there isn't a functioning cultured meat production facility out there to study, a place where an observer can conduct research backed with data about the realities of growing meat.

In December 2018, some of the world's leading researchers of cell-cultured meat converged in Holland, meeting at La Bonbonnière restaurant in Maastricht for a two-and-a-half-day conference organized by Mosa Meat's Mark Post. They discussed the usual topics: the development of liquid medium, bioreactors, tissue structures, cell selection, sociological trends around cultured meats, and the latest research into the sustainability of cell-culturing meat on a large scale. Researchers came from around the world, including the United States, Finland, Japan, Spain, and England. Hanna Tuomisto, a University of Helsinki researcher and lead author of a much-cited 2011 Oxford University study on cultured meat sustainability, presented some of her latest, unpublished findings.

Her work showed that total energy consumption by cell-culturing meat would be about four times higher than previously shown in the 2011 study—not news the cell-cultured meat companies were hoping to hear.

"Many of the start-ups got quite scared that we would publish these results," Tuomisto told me. "We were planning to submit the paper, but then we decided not to publish because we understood the worry."

The concern was that the media would seize upon the difference between the 2011 study and her unpublished research with little regard for nuance, that it would be a "big story" that would distort Tuomisto's findings and potentially harm the effort to get cell-cultured meat to market. So what was the nuance here?

Measuring the life cycle analysis of any food process—ascertaining its environmental impacts from every stage of production—is a complex scientific and mathematical challenge. For Tuomisto, the devil in the details was the liquid medium. Ultimately, as she explained to me, her 2011 Oxford University study theorized data under the assumption that the companies would wind up using a cyanobacteria-based medium. That's one of the most environmentally friendly mediums she could have picked to plug into her equations when coming to her findings.

There's just one glaring problem: To her knowledge, none of the companies are actually using a cyanobacteria-based medium. She's asked them for specific details about their processes and what materials they're using, but they have, characteristically, held such sensitive data close to their chests. So for her most recent work she operated on the assumption that the companies were using fetal bovine serum, something she knew for certain the companies had used before in their processes. A lot of the companies, including JUST and Memphis Meats, have said they've moved beyond such controversial and expensive serums—but they've not publicly shared what they are using as a substitute. The start-ups' need to guard their intellectual property makes presenting totally accurate research almost impossible for academics such as Tuomisto, and for reasons beyond her control. It highlights a need for greater transparency.

"They don't say what they are doing and they don't present at the conferences, so I'm not completely able to track the progress," she said, pausing a moment. "I think they are too optimistic."

Again, while there are a lot of reasons to be excited about the environmental promise of cultured meat, we have no idea how sustainable these company processes actually are, so any marketing that casts their product as an environmental godsend should be taken with a grain of salt. Still, it is generally agreed upon by people in the field and in academia that, compared with conventional meat production, cell-cultured meat's biggest energy savings will come from land and water conservation.

Regardless, the work happening inside JUST's and other companies' laboratories is still exciting, the first tiles on what may well become a mosaic of new food possibilities; a convergence of ingenuity and science that takes the ideas van Eelen worked so hard to pioneer well beyond science fiction and into a future filled with possibility.

In September 2017, Tetrick announced that JUST had acquired van Eelen's original patents, along with similar patents held by Jon Vein, the Los Angeles business titan. In the same announcement, the company said it was appointing van Eelen's daughter, Ira, as an official adviser to the company.

The news gave other cell-cultured meat start-ups in the field pause, as they were taken aback by a comment made by Wayne Szeto, who oversaw JUST's intellectual property. Szeto told me, for *Quartz*, "We're open to having discussions around licensing. This is not going to be 'play with us or don't play at all,' not in that black-and-white way."

But it remained unclear exactly how JUST could or intended to use the patents in any legal sense. To date, the company has not used them to lay claim to a specific production process.

In Tetrick's view, he'd elected to carry some of van Eelen's burden, eager to fulfill a legacy more than sixty years in the making and to be the first person to make a cell-cultured meat sale. Such an event would herald a new chapter for the field, opening the door for a novel idea to capture public imagination. Whether he was ready for such a role, and just where he'd be able to carry it out, however, was entirely unclear.

4

THE DUTCH
DREAM

On December 29, 2017, just two days before New Year's Eve, Tetrick and his golden retriever, Elie, boarded a nonstop KLM Royal Dutch Airlines flight in San Francisco that would take him halfway around the world to Amsterdam, where he hoped to make the first commercial sale of cell-cultured meat.

The journey had monumental potential. For several weeks prior, he and a small team at JUST had toiled over every minute detail of the trip, an intense effort to maximize the chances this rarefied mission would be clocked as a success. They researched shipping rules and times and even designed a nutrition facts panel for the box in an attempt to be overtly transparent about what was inside.

There was just one complication. To Tetrick's knowledge, no

one had ever shipped cell-cultured meat across the world and into a totally separate consumer marketplace. Just how a package of cell-cultured meat would get through customs—if at all—remained a mystery. But it was imperative that it did. Tetrick was operating against a tight deadline.

On January 1, 2018, the European Union would implement an update to its 1997 novel foods legislation that would expand the definition of nontraditional foods to include products "originating from plants, animals, microorganisms, cell cultures, minerals, etc." as well as those that were made using new and innovative production processes. Under these rules, a company such as JUST would have to fill out a formal application and submit to the government a request to have its cell-cultured duck chorizo considered, tested, and approved to be listed as a novel food. That update to the EU's food policy was all fine, in Tetrick's estimation. The issue was that the process would likely take up to two years for bureaucrats to complete once the company submitted an application. With that in mind, Tetrick wanted to make an effort to sell his product into the market before the new law took effect. If his cultured meat was sold before the new regulatory regime, he felt there would be a legal basis for the argument that JUST's cell-cultured meat could be grandfathered into the food system. He'd still have to apply to sell a novel food, but as he understood it his cell-cultured product could be sold and consumed while the application was being processed. It was a long shot, but he was optimistic about his chances.

There are material aspects to making cell-cultured meat a

reality—nailing the science, raising capital, pitching products to retailers and restaurants—and then there are more amorphous elements: building a good story around the technology.

For Tetrick, launching JUST's cell-cultured meat in the Netherlands was a narrative gold mine. It was the birthplace of the concept, the nation has a history of progressive climate policies, and he owned Willem van Eelen's patents.

He had roughly forty-eight hours to make that story work. The clock was ticking.

He arrived in the Netherlands at the end of a ten-and-a-half-hour trip that left him standing in Amsterdam's Schiphol Airport arrivals hall, waiting for his package to spit off the airliner and down a cargo carousel. Minutes passed—long, excruciating minutes. He waited. And waited. Once Tetrick realized he was the last person standing at the carousel, he dashed to a nearby customer service counter. Nightmare scenarios played out in his head, not least of all the possibility that two pounds of priceless cell-cultured meat had been plucked from the arriving cargo by a team of skeptical customs personnel.

As the minutes ticked by, Tetrick nervously prepared himself for a showdown with Dutch customs officials.

An airline representative checked one of the company systems on a computer.

Nothing.

Another representative checked a separate system, scrolling through lists of package identification information. A minute passed. Then another long minute.

How many systems can they possibly have? Tetrick wondered.

One of the agents looked up.

"Well, it looks like it didn't get off the ground in San Francisco," he said.

It never left the United States. It was, in fact, still sitting more than 5,400 miles away, thanks to the airline's own shipping error.

On one hand this was good news—the meat hadn't disappeared into a Boschian customs nightmare. But the time line now felt impossible. Was there even a shred of a chance it could be shipped around the world in time to meet Tetrick's January 1 deadline?

He pleaded with apologetic KLM representatives, who assured him the package had been located and would be on a flight to the Netherlands as soon as possible. In the meantime, he'd just have to wait. By then it was late afternoon on December 29. The package had approximately thirty-six hours to get from the United States to the Netherlands, through customs, and into his hands—all during the busy holiday season—if he hoped to make a sale before the new food law went into effect.

Tetrick had planned to stay with Ira van Eelen on her spacious houseboat. When the two met, their anxiety dampened any excitement about their shared, potentially historic moment. Both their fates were dependent on an international airline.

She was nervous. So was he. To kill time he struck out into Amsterdam on his own, hoping to find some way to distract himself. His wandering took him to a cinema, where he watched half of *Molly's Game* before walking out of the movie theater. Even the true story of an Olympic skier's transformation into a poker kingpin failed to harness his focus. He called Ira and she drove to the movie theater to pick him up. They returned to the houseboat and he fidgeted until it was time to sleep.

When he woke up the next morning, Ira was already up and preparing breakfast for her guest. She stood over a stove, sizzling up sausages and bacon from a local vegan butcher shop. The pair ate their meal in relative silence, munching on vegan "meat" while waiting for a delivery of higher-tech vegan meat.

Then there was a knock at the door.

A courier handed over a box. *The* box. KLM hand-delivered the package, fresh off an overnight flight from San Francisco. And just like that, Tetrick had a little more than fifteen hours to make the historic sale. The box contained one package of ground duck chorizo and a single duck sausage link.

With the meat in hand, Tetrick and Ira piled into her van and drove north of Amsterdam and into Zaandam, a city along the North Sea canal and a major Dutch industrial hub. They were en route to a restaurant owned, in part, by a man named Paul Riteco. The name of Riteco's restaurant was De Hoop op d'Swarte Walvis, which means "Hope for the Black Whale." It's a historic spot, but the white-tablecloth restaurant of modern times is a far cry from what the place used to be. Back when the Dutch still harvested whales for meat and oil, boats returning from hunts at sea would drag the massive carcasses up to the current location of De Hoop op d'Swarte Walvis to strip and process the whales. The area reeked, especially in the summertime, when the smell of the carcasses heated by the sun would waft inland through the streets.

Zaandam itself is an idyllic, mostly man-made stretch of country, marked by its ruler-straight canals, historic wooden houses, and collection of traditional windmills. The primary reason Amsterdam ascended to such prominence in the Netherlands was because of the invention of the sawmill—which ushered the country into its industrial golden age. In the sixteenth and seventeenth centuries, Amsterdam's shipbuilding industry was an important source of economic activity for the country, thanks in part to the rise, expansion, and ambitious global conquests of the Dutch East India Company. The advent of the sawmill made the construction of big wooden ships faster and cheaper than the kinds being made by hand. As with many new technologies, it ruffled the feathers of existing craftsmen, upsetting in particular the sawyer guild in Amsterdam, which successfully pushed for a city law that banned wood sawmills from opening within city limits. Looking to the south and seeing an opportunity, the peasant community several miles away in Zaandam decided to embrace the new technology for itself, and in doing so shifted the center of Dutch shipbuilding from Amsterdam to Zaandam.

The wood mills were soon joined by flour, oil, paint, and paper mills—all of which helped make Zaandam an industrial leader in the Netherlands and a place of interest around the globe. At the time, the small European country was among the most developed in the world. Indeed, the Russian czar Peter the Great traveled from St. Petersburg to Zaandam himself in 1697 to spend the better part of a week studying the craft of Dutch shipbuilding, hoping he might take those skills back to his own country, part of an attempt to usher in a period of modernization for his people.

Because of Zaandam's rich industrial history—and the city's willingness to embrace the benefits of new technology—Ira felt it would be an excellent spot to launch the new cell-cultured meat. It was a move that she thought would be monumental not just for her country, but also for her father's legacy.

Today Dutch tourism agencies market Zaandam as a place to see life in the Netherlands as it used to be. With its well-preserved wooden homes and, of course, the windmills, it's a spot that draws tourists by the thousands every year. For that reason, Ira and Tetrick had envisioned it as the ideal location to introduce cell-cultured meat to the world—a convergence of old and new before the eyes of inquisitive global visitors. And what better than a location that once existed, in large part, because of the death of animals.

"Millions and millions of tourists come here and we want to feed them JUST meat," Ira told me.

Tetrick made the sale on December 31. Riteco bought a little more than a pound at the very-discounted price of $11 per pound. Riteco's young daughter was present at the sale, hovering over the table next to her father, who'd brought her along in hopes of giving her a rare glimpse into what he hoped would be a big part of her future. Tetrick said he also intended to set up a miniature production facility so that the restaurant might produce its own meat to feed consumers.

A tiny amount of meat was also sold to the NEMO Science Museum in Amsterdam, which was planning an exhibit around the concept. In video footage of the sale, Tetrick is sandwiched between Ira and the director of NEMO. To his left, Ira is quiet, rosy, and stifling tears. Tetrick notices, and asks her what her father would think if he were there to see it. She rocks back and forth and smiles, seemingly unable to get a word out.

"Yes. Yes," she replies in a near whisper, this time unable to withhold her emotions.

For that moment, Tetrick had made history. On his return flight home, he basked in the glow of sidling up to the edge of history and taking a step forward into the unknown.

But what he didn't know was that his victory would be extremely short-lived. The regulatory landscape in the EU is as fickle as—if not more so than—that in the United States.

5

PANIC IN
AMSTERDAM

A black van rolled to a stop on one of Amsterdam's sleepy east end streets, bearing news on a misty night.

I opened the door and climbed into the passenger seat. Ira van Eelen wasted no time switching gears, lurching the van forward and down a slender avenue flanked by houseboats anchored in a quiet canal on our left, and tall Dutch-style apartments standing like sentinels on our right. The honeyed veil of streetlamps illuminated her grim expression.

"I've got some bad news," she said.

Amsterdam is, among many things, perhaps one of Europe's most self-determined cities, rapidly built by a population that possesses what city historian Russell Shorto has called "an ingrained commitment to openness." I'm here to learn more about the history of cell-cultured meat, which included learning about Ira's father's history. But this wasn't a happy encounter. Sitting behind the wheel of her van that damp evening in March 2018, Ira looked anxious and angry, if not a little defeated. It wasn't her proudest moment as a Dutch national.

She nudged the accelerator, speeding us north through the Oosterdok tunnel, driving directly beneath the nationally acclaimed NEMO Science Museum where, just hours earlier, on that chilly afternoon in March 2018, Dutch officials had walked unannounced into a corner office and delivered a stern warning to one of the museum's managing directors. They were representatives of the Netherlands' food safety agency, and they were alarmed that just a few months prior the museum had been one of the Dutch entities—the first in history—to purchase high-tech, cell-cultured meat from JUST. The meat had yet to receive approval by any world government, and even though the museum sought only to display it in a forthcoming exhibit, government regulators had decided to crack down on the sale, calling it an illegal purchase of illegal goods.

It was a sobering development for everyone involved, but Ira was particularly ashamed that her government would be the first to take an aggressive stance toward one of the most innovative and potentially impactful future foods ever conceived. It was also an early shot across the bow of this new wave of start-up companies working toward making cell-cultured meat a reality—a clear sign that the pathway to market would be a slow and arduous one as governments looked at their products with skepticism.

Even more slow and arduous than impatient Silicon Valley entrepreneurs expected.

Ira's van rumbled through Amsterdam's city center where her friend Koert van Mensvoort was waiting to be picked up outside a bus station. Aside from being an enthusiastic supporter of the idea of cell-cultured meat, van Mensvoort is also one of the authors of a cookbook that details conceptual, wacky, and absurdist recipes that would be made possible by cell-culture technology. There's "maple-smoked labchops," which are reimagined as a meat product presented as a log that can be sliced into circular cuts of cell-cultured lamb meat. The book also includes "roast raptor," which makes the case that with the right cells humans could one day taste some of the Earth's prehistoric dinosaur meats. One particularly macabre illustration showed an arrangement of whiskey-soaked meat cubes, each pierced with a little toothpick. Affixed to each toothpick were photos of figures in pop culture, including Kanye West and Miley Cyrus. And written on the page next to them was a description of how a person could appreciate celebrity culture in a whole new way—by eating a minuscule version of their likeness.

On this night, his jovial demeanor contrasted starkly with Ira's. She tried in vain to hold back tears as she stared straight ahead with both hands affixed to the steering wheel, pushing us further into the night and across Amsterdam's city limits.

"They went into NEMO to accuse NEMO of purchasing illegal meat," she explained to the two of us in the car, bitter tendrils wrapped around her words.

"Ah, really? So they really want to fight," Koert replied, a comment that hung suspended for a moment across a yawning silence.

"They were not very friendly," she said. "I'm really very sad that even still today they take offense to this."

She pulled her van back onto a darkened Dutch highway, pointing us north toward Zaandam, toward the historic restaurant at which Tetrick sold some of his meat to Paul Riteco. The road slopes downward, dipping beneath the NEMO museum itself, which stands above the roadway creating a colossal underpass. Ira pointed to a corner office at the building's exterior, the spot where Tetrick had made the sale happen just weeks before.

"For me this is so hard to take, because in my father's view, and in my view, we are ready to do this," she explained. "The Netherlands is losing this opportunity. I felt like this really happy person, that I had found [the place] that could actually and finally make this stuff."

Koert chimed in optimistically.

"I see it as just another hurdle, and it will be taken and slow things down," he said. "But I think it's something to be proud of."

We pulled off the road and onto a gravel lot. Ira eased her van into a narrow parking slot and cut the engine. The three of us stepped into the murky night, engulfed by a March chill. I zipped my coat and tightened my scarf as we made our way down a winding lane, toward the dark bank of the Zaan River where, stationed along the water, several traditional Dutch windmills towered overhead ominously. Sturdy watchmen in the night, they were illuminated only by a faint wash of moonlight. Apart from our own footsteps, the evening was quiet, almost funereal. Ira sniffled. You could see our breaths in the air.

She looked out into the dark expanse of the Zaan, the currents whipping the river southward toward Amsterdam's waterfront. Ira motioned toward a wooden structure a far stone's throw away. Its many windows were illuminated, emanating a warm and inviting glow. It was the restaurant, De Hoop op d'Swarte Walvis.

We were scheduled to have dinner at the restaurant, but the sight of it relit the anger Ira was feeling toward the Dutch government. The softness in her voice evaporated. With everything at stake—the changing climate, the Paris Climate Accord going into effect in less than a year, and billions of animals dying every year unnecessarily—how could her own government have been the first to strike down an attempt to get cell-cultured meat to market? If the government had been supportive, it would have sent a powerful message to the rest of the world, Ira said.

"To show them that we were a great nation, and that we still are a great nation," she said, pushing a bit of gravel aside with her shoe.

On a continent where culinary traditions have helped shape the law of the land, where countries are fervent protectors of their gastronomical heritage, getting new, high-tech meat products into the market was never going to be easy. The EU Commission plays a role in deciding how these products are released, and that includes just what they are called across all its twenty-seven member states. Even at the time Dutch food authorities put a halt to Tetrick's plan, the environment in Europe around newfangled versions of traditional foods was already the subject of heated debate.

In 2017, the German agriculture minister Christian Schmidt called for an all-out ban on terms such as "vegan currywurst," a plant-based version of pork sausage. He argued that such non-meat substitutes confuse consumers. Those types of squabbles would only intensify in subsequent years, and answers to them would always have to be worked out in the halls of the government in Brussels.

We began making our way to the restaurant. Everything about the place, with its fancy white tablecloths and rustic cross-beamed ceiling, was imbued in Dutch history, from the food on its menu to the priceless blue-and-white Delft tiles cemented to its walls.

Owner Paul Riteco met us at the front door. He was already aware of Ira's bad news and the fact that it officially prevented him from being the first restaurateur on the planet to sell and serve cell-cultured meat to a consumer. He walked us through the halls of the building, sharing the long history of the place. Above me hung a long and deadly relic of the past, a harpoon used to catch and slaughter whales.

The mood at the dinner table was just as dark and deflated. Ira, Koert, and Paul constructed and reconstructed the events of the day for themselves. Had food authorities all shown up at the restaurant? Not yet. Would they come for the cultured meat Riteco had stored in a freezer? He wouldn't let them.

It was the sort of angst that comes when a few people are faced with the insurmountable authority of government—with all its opaque systems and bureaucratic inner workings. Who, exactly, had made the call to storm into NEMO? What person needed to be contacted? What would be the best way to change the government's mind?

Two options seemed to emerge. Either they played the government's game, what would surely be a years-long effort to engage, insist, and plead with regulators to take action for a greater good. Or, as Ira was planning in a fit of anger, there was always civil disobedience.

Together they brainstormed potential strategies for moving forward. Maybe they'd protest. Maybe they'd launch a public awareness campaign. Hell, maybe Ira would illegally grow meat in her own home, a flagrant disregard for the rule of law.

"I am the daughter of Willem van Eelen," she said. "I think the whole world will understand my civil disobedience."

Maybe it wouldn't, though.

Months later, on April 1, 2019, the European Union parliament's agriculture committee would vote to affirm a measure that would prohibit the makers of vegetarian and vegan meat and dairy food alternatives from using marketing terms such as "burger," "steak," and "milk." If the makers of cell-cultured meat were taking the temperature of reception to their products in Europe, it was decidedly chilly.

And while it was still too early to tell just how any of this would impact the argument that cell-cultured meat is "the real thing" on a molecular level, the political movement in the EU among farmers hasn't shown many signs of easing up. It could be decried as protectionist, but the status quo is powerful. It's a challenge for vegan and vegetarian upstarts looking to capture swaths of the meat market: If they want to compete, they're going to have to be more creative.

But what if JUST simply went through the motions of officially applying to be served up as a specialty food, I asked. After all, couldn't Tetrick still aim to sell his meat on Dutch soil in the near future, once the European Union's necessary regulatory conditions were met?

"Josh is not going to do the application," Ira replied, glumly.

It was a missed opportunity for the Netherlands, she repeated. She knew that Tetrick prioritized getting the meat to the market sooner than later, and that meant turning his energies elsewhere, letting go of a Dutch dream and postponing any intention to engage in an application process that promised to take at least two years to finalize.

I mulled over the conversations I had with Ira. The fact that Dutch authorities abruptly and abrasively put a stop to the effort to sell and serve cell-cultured meat first was an emotional blow for her. She was in the tough spot of trying to figure out whether to let go of her father's legacy or to double down and find a way to make a difference by another route.

A few days later, after leaving Amsterdam and heading back to the United States, I picked up the phone to check in on her. To my surprise, when she answered, I detected a singsong levity in her voice. As it turns out, I caught her just around the time she was beginning to hatch a plan to have herself thrown in jail, part of an elaborate stunt to draw attention to the issue of cell-cultured meat. She was willing, even eager, to be arrested, she said, for illegally growing cell-cultured meat in her own kitchen. She had even reached out to Japanese cell-cultured meat company IntegriCulture about obtaining one of their novel microwave-size machines used by children in Japan to grow very primordial and totally unrecognizable forms of cell-cultured meat.

It was all about the act of defiance, she said.

I was taken aback at the idea, though not because of her enthusiasm. No, her commitment to the cause was clear. But she was resorting to the type of tactics that absolutist vegans employ, a far cry from the way a vegan entrepreneur such as Tetrick seeks to create change. He was more interested in pragmatic approaches. Perhaps she felt abandoned by pragmatism.

"I think Josh has already forgotten about the Netherlands and doesn't think anything is going to happen here," Ira said, her voice saturated with disappointment.

A few weeks later I was with Tetrick at JUST's headquarters in San Francisco. Indeed, he had moved beyond the collapse of the Dutch deal. His explanation for why the effort fell apart was matter-of-fact. It was already ancient history.

"They decided, unequivocally, that cell-cultured meat is under the umbrella of the novel food regime, which basically means, in their parlance, that if it wasn't sold before 1997, it's considered a novel food," he said, adding that being grandfathered into the Dutch food system likely wouldn't have worked, even if the meat was allowed to be served in that historic restaurant on the bank of the Zaan River.

His mind was already elsewhere. Amsterdam was behind him. He'd met with regulatory powers in Singapore—and had even begun communicating with local restaurants within the Asian city-state—to see if the same idea might land there. Clearly, Tetrick had abandoned his effort to sell his cell-cultured meat in Willem van Eelen's home country first. It was a tough call, but one based purely on business and expediency.

Meanwhile, JUST focused the bulk of its cell-cultured meat attention on other parts of the world, casting a wide net. One member of his team, spokesman Andrew Noyes, had even tossed out a fleeting idea of serving up the company's cell-cultured meat on a boat somewhere in international waters, where national rules around food don't apply.

Tetrick had hoped to have cell-cultured meat on the market in the beginning of 2018. With that plan having fallen through, and a fuller scope of the regulatory hurdles coming into clearer view, his window of opportunity for having it out by the end of the year was also closing fast. It was still unclear how the idea would play among American regulators, and he was looking for potential interest by engaging with governments in Asia, including Singapore, Hong Kong, and the United Arab Emirates.

The research and development of cell-cultured meat was certainly a major challenge for any entrant into the small field, but, if anything, staring down the challenges posed by international regulators would truly prove their mettle.

6

UNTETHERED

He remembers how he tightened his fingers around the grip of the baseball bat, how its barrel floated ever so slightly over his teenage shoulders. Tetrick summoned the energy to push it through the air with powerful gusto.

Swing!

Back to position.

Swing!

The bat whooshed through the air and rotated wildly back behind him, where a young classmate about his same size stood watching. Josh Balk's ginger hair accented a boyish smile that lingers even to this day. He stepped forward to tweak Tetrick's form.

Nothing has ever come easily for Tetrick. He grew up in a chaotic household in Birmingham, Alabama. His mother worked as a hairdresser and his father held a mishmash of odd jobs. They often couldn't pay their rent, resulting in constant upheaval. By the time he was thirteen, the family had moved fourteen times. Each time, they dealt with new homes, new friendships, new hardships.

Tetrick's parents divorced when he was just nine years old. Four years later the family even moved together to Wayne, Pennsylvania, a suburb of Philadelphia more than 870 miles north of Birmingham.

"Mom and Dad moved together just because they wanted to keep the family intact as much as they could," Tetrick recalled.

As the new kid in school, he congregated where the others were. The school bell would sound at the end of a long day of classes, and the halls of Radnor High School suddenly brimmed with the anxious, kinetic energy of several hundred jostling teenagers. They bounced from their lockers to the three front doors of the building, flooding out into the sprawling suburban scenery that surrounds Philadelphia.

He was one of the many who hastily made their way to a grassy area near the main building, where big yellow school buses belching clouds of diesel fumes would rumble through and stop to pick them up. Nearby was a busy baseball diamond. As students waited for their rides, they would congregate near home plate, killing time with wide practice swings.

That was the way he met Balk, when a bat found its way to his own hands. And that's all it took. A mutual love of sports and a friendly gesture set the foundation of a friendship that would last deep into both men's adulthood. At the time, it was exactly what Tetrick needed. He aspired to be a linebacker for the Eagles. Balk was determined to pitch for the Chicago Cubs. The Balk family St. Bernard was actually named Héctor Villanueva, after the Cubs' backup catcher and first baseman.

There's a photo of Tetrick and Balk from their high school days that portrays the bond the boys shared, an artifact that shows how the friendship between the pair evolved. Tetrick looks concerned. He's wearing a gray muscle tank top that exposes well-built upper arms, and he appears to be shouting into a bronze megaphone toward circus tents being erected well outside the borders of the image. But it's behind him where the real action is happening. Back there, just barely out of focus, Balk looms, smiling so broadly and so knowingly that I can't help but suspect he'd somehow masterminded the entire scenario.

I wasn't far off the mark.

"This picture here, I would say, sums it all up," Tetrick told me. "This is me, though obviously a different physical person."

He taps on Balk's head.

"But this smile here, he's basically convinced me to do this," he says. "He basically had somehow gotten in my mind and my heart and was like, 'Okay, convince people to stay away from the circus here.' And he's just smiling. He's like, 'I've organized this whole thing. I strategized it.'"

How would Tetrick sum himself up in a word?

Untethered.

That's the word that came to his mind when I asked him. It's not necessarily what you want to hear about the CEO of a billion-dollar company. But Tetrick's sense of his own aimlessness, and his search for anchors, is the reason he first bumped into veganism and why he decided to be an activist. It's the reason he wound up starting a company. And it was behind his decision to use his company not just to focus on making plant-based foods, but also to charge into the field of cell-cultured meats.

But before someone can fully appreciate the specifics of that plan—the one to introduce to the world an unexplored new future for meat—they must first consider what led Tetrick to that rarefied position in the first place. Because, truly, much of it had very little to do with him at all.

Josh Balk came into Josh Tetrick's life at a moment in which he most needed a comrade—someone not only to accompany him through the tumult of adolescence, but also a person in whom to place his faith. Someone to look up to and admire, to help him believe that he himself might have a role to play in this world that is inherently larger than any one person. They initially connected through sports, but as their friendship evolved, so too did the nature of their conversations. Tetrick has always been an animal lover, but whether he would have connected those dots to the animal liberation movement on his own, let alone have the moxie to take up the cause as an activist, is debatable.

That gray afternoon from his youth, protesting the circus, might have been one of Tetrick's first forays into the animal liberation movement, but perhaps more important, it represented a moment in his impressionable youth that gave him a sense of belonging. Balk was his surrogate brother. Tetrick estimates that he spent more time at Balk's house during high school than he spent at his own.

They were seemingly attached. They played football and baseball together, and had long conversations about purpose. They'd sometimes venture to the local Hollywood Video store around the corner from Balk's house to rent wrestling videos and worn-out tapes of old NBA games. That's where they first ran across the tape that inspired them to take up the cause of advocating on behalf of animals.

"I think it was just sitting there," Tetrick recalls. "We were like, 'What is this?'"

A peppery static-specked scene morphed into an intense set of visuals. There's a person lying on a surgical table at the center of a clinical operating theater. Surrounding this person is a team of doctors, all of them gowned and masked. The camera zooms in on a human heart; red, glistening against the fluorescent lighting and pumping at a harried pace as the surgeons poke and prod around it. It quickens. Then, suddenly, the organ gives a final sigh before collapsing into a dull stillness, accompanied by a high-pitched beeping and the slow continuous whine of an accompanying piece of hospital equipment. The screen fades to black. A bright crimson font appears, revealing the title of this obscure and seemingly perverse 1978 mondo horror film: *Faces of Death*.

The man who created it went by the pseudonym Conan Le-Cilaire, a name that he told *The Guardian* he picked because it meant "Conan the Killer" in French. When it was released on VHS for home viewers, the film was especially attractive fodder for curious teens, enticed by the package's boast that it was "BANNED in 46 countries!" In reality, it was banned in only a few. Even so, it lived up to its shock value, and many politically active vegans today talk about how the movie influenced them to become activists.

The screen switches back to imagery of human bodies, this time in a morgue, fingers and toes twisted like thin and gnarly tree branches, contorted by the natural whims of atrophy. A mortician begins slicing into one of the bodies and I quickly reach forward in the dark to fast-forward ten minutes into the film. Now I'm greeted by two pit bulls in a dogfight.

"These animals only know one way of life, they have been conditioned by men to make war on their own kind," a narrator says. Later, scenes of sheep and cows grazing in bucolic fields transition into footage of sheep being shocked to death with electrical prods and cattle being sliced at the throat.

"Western culture has developed the ultimate killing machine to feed the hungry masses," the narrator continues. "It's called a slaughterhouse."

For young teenagers sitting in front of the television screen, this was a glimpse into the macabre reality of death, exposing a very real part of life that we're typically sheltered from witnessing.

It left an imprint on the two boys. As they put it today, the idea of death was an abstraction. To see death in such stark relief, retreating from life in such gruesome and violent detail, ripped away the mystique that often shrouds its sheer physical- ity. And for Tetrick and Balk, who were on the verge of exploring veganism at the time, it posed a single, obvious question:

"Why does something have to die for us to live?"

After high school, Balk went to Keystone College, a private school just northeast of Scranton, Pennsylvania. He graduated and moved to Washington, DC, where, while walking down a street, he happened by the national headquarters for the Hu- mane Society of the United States. He walked inside thinking he'd find animals, but quickly learned it was more a hub for peo- ple in suits, many of them lawyers, who worked toward protect- ing animals. He inquired about the work being done there and quickly found himself working as an intern in the government affairs division. He's been there ever since, most recently work- ing to try to convince some of the world's largest food companies to switch to cage-free eggs.

By contrast, Tetrick chose to go to West Virginia University, where he spent two and a half years before transferring to Cor- nell University in Ithaca, New York. After receiving his BA in 2004, he was accepted to University of Michigan Law School. It's not that he felt particularly passionate about studying law— he still had no idea what he wanted to do with his life—but he said he figured taking the LSAT and going to law school couldn't hurt. Maybe it would give him skills that could be used to be a force for positive change in the world, he thought.

"While I was in Michigan Law School I spent time in Kenya,"

he said. "My law school hated me, it was a frustrating time for them because I basically realized in law school that I didn't want to be in law school."

He got his law degree, though, and headed back to Africa, working a few years for a United Nations project with the Liberian government to help the country identify businesses that could help solve social and environmental problems and create a system of tax incentives to try to lure them to the country and invest their capital. He would also spend time in South Africa and Kenya. While in Africa, Tetrick worked on social justice projects, including an effort to get child prostitutes off the streets and into schools. He was briefly involved with More Than Me, an educational charity that, well after Tetrick left it, would be the subject of much criticism as its cofounder was accused of abusing pupils at schools.

Even at its best, it was abstract work. Though Tetrick believed what he was doing had value, it wasn't the kind of hands-on approach he desired. Comparing himself to Balk—always his lodestar—Tetrick realized how unfulfilling his own work truly was. While he was toiling through the bureaucratic complexities of the UN and economic policy, Balk was working as an animal activist with a concrete goal of trying to convince corporate food companies to adopt more animal-friendly practices, including switching from caged-hen eggs to cage-free eggs. Not only was he achieving real results, he was increasingly seen by his peers as a rising figure within the movement.

"Josh was clearly using his life to do something positive, and I had spent this time in Africa searching for something to do that was positive—something that we could all measure, something that I could see," Tetrick said. "It was all things that on paper felt nice. But for me, the experience with nonprofits, the experience with international institutions wasn't doing it."

Tetrick came back to the United States in 2009, taking a job at a firm in Richmond, Virginia, called McGuireWoods. He had his own office and an executive assistant, but in spite of the perks, even on his first day he felt like a square peg being shoved into a circular hole. His assistant showed him to his office and shut the door. Tetrick immediately felt claustrophobic.

He opened the door and made a beeline toward the desk where his assistant sat. Walking over, he asked, "Can I take this work to a coffee shop that's, like, a block away or so?"

The assistant peered up at him a moment, then replied matter-of-factly, "No. This is your first day. You can't just take all this to a coffee shop."

He went back to his office, closed the door, and immediately felt the full weight of his decision to work at the firm.

"Even if that work was objectively good for the world—and it was objectively not good for the world, but even if it was—I couldn't do it. I can't handle that. The prism was too narrow in how things are approached. I'm sure I was a shitty attorney there."

Cloistering himself in an office at a prestigious Virginia law firm had obviously been a misstep.

The firm soon realized that, too. In March 2009, he penned

an op-ed for the *Richmond Times-Dispatch* entitled "You Can Save the Planet," including one sentence about some 70 billion animals suffering from "cruel and inhumane treatment inside factory farm walls."

There was just one problem: McGuireWoods represented Smithfield Foods, a multinational meat-processing company that, as of 2006, was processing some 27 million hogs each year, accounting for 6 billion pounds of pork. Tetrick was fired shortly thereafter. He was happy to leave.

He kept in touch with Balk the entire time, brainstorming with his friend about what role he might slip into that would help him find some respite and alleviate the anxiety of feeling unmoored.

"I think he sensed in me that there was potential, but there wasn't a vessel for the potential," Tetrick said. "When he first met me there was a vessel, it was called the baseball field, you know? Or it was the football field. But then, when we got a little bit older, he knew I didn't want to work at the Humane Society, so what do you do with me? Where the fuck do you put this guy, you know?"

Tetrick was still deeply interested in animal liberation and activism. The problem, though, was that he'd come off several years spent in Liberia, South Africa, and Kenya, watching up close how the nonprofit sector repeatedly proved ineffective at solving some of the problems it was supposed to be addressing.

"We were always raised with this false choice: You can either work for a nonprofit and do a lot of good and not make any money, or you can work for a company and not do a lot of good, but maybe you can donate," he said.

He thought to himself: Maybe he should truly embrace capitalism.

"It's just more me," he said.

And so, more than a decade after high school, he found himself crashing on a sofa that wasn't his, in a Los Angeles house that wasn't his, in a city that wasn't his. Here he was plodding through the tail end of his twenties, armed with an impressive résumé, and yet he was untethered, wandering through 2011, trapped in a fog and wondering what he should do with his life. He was staying in the house of his ex-girlfriend and friend from college, Jill Hundenski, who remains close to this day.

In her sunny living room, he'd sit with his dog and rack his brain over what to do next. On paper he was absolutely accomplished, but on the inside nothing added up to something truly meaningful to him. He was still left sitting on a friend's sofa with no direction and no idea what should come next.

"It wasn't a financial rough patch necessarily, it was more, 'What the fuck am I doing with my life?'" he told me. "I tried to do what I did in Africa and it didn't work out for me as I'd expected. I tried to do a handful of these things that I thought mattered."

By happenstance, Balk, who by this time was a vice president at the Humane Society, was in the room one day, too, stopping by while on a business trip to sunny California.

"Josh, I'm trying to figure out what I want to do," Tetrick said.

Tetrick's childhood friend had become the picture of success in his field. He'd worked tirelessly to get people to switch from caged-hen eggs in a larger effort to improve the lives of hens. Balk had worked with companies including Walmart, General Mills, Kroger, and McDonald's to change their supply chains. He'd worked on statewide ballot initiatives to create political momentum and legislative change. As a result, he had earned the respect of animal advocates.

"Listen," Balk said to his struggling friend. "You really need to start a company. We really need to start a company that finds a way to hammer one part of industrial agriculture. What if we start with eggs?"

In hindsight, Tetrick realized the irony in Balk zealously advocating starting a company. After all, Balk had never been particularly entrepreneurial. He'd certainly never founded a successful business and he didn't feel very corporate. He was very much a nonprofit guy. And yet Balk was suddenly embracing this idea that capitalism, in the hands of vegans, might actually be used as a powerful tool against the market monoliths the two men sought to tear down.

The idea wasn't foreign to Tetrick at all, in fact it had been invigorating him for years. In that fateful op-ed he wrote for the *Richmond Times-Dispatch*, he expressed as much when he said, "The world's biggest needs align with dynamic opportunities to engage your strengths, find meaning—and make money."

On December 11, 2011, the two men cofounded JUST. It operated out of a garage at first, and its initial products would be a line of vegan condiments, including eggless mayonnaise. At the time, the primary goal of the company was to try to make a dent in the egg industry, with the ultimate hope of creating a convincing plant-based liquid egg product.

From the beginning, Balk was hands-off. Today he has zero equity in it at all, though he's intimately aware of what goes on inside JUST through his near-constant communication with Tetrick, who from the get-go was JUST's chief executive officer.

Looking back, Tetrick credits Balk entirely for the creation of JUST.

"If not for him, the company wouldn't have started," Tetrick says. "He had the initial idea, the concept, pushing me, all of it," he said.

Tetrick did what a lot of Silicon Valley's nervous first-time entrepreneurs do. He assembled a small team to begin working on early-stage products and spent his own time wooing potential investors for seed money to help his new business grow.

And for a while it was enough. The company's work and mission elicited a lot of attention from Silicon Valley venture capital firms.

The company would ultimately attract notice—and some measure of notoriety—for its suite of vegan condiments and cookie dough. But in those earliest days the primary focus was creating a vegan egg replacement. The proposition carried extra oomph because it coincided with an increasing amount of public education and activism around the caged-hen egg industry (helped along by Balk's work). The question Tetrick posed to potential investors was simple and straightforward: If you could eat a convincing egg product that's as cheap and good as the real thing without harming an animal, wouldn't people embrace it? He believed it would work, and he convinced the right people that he was the person to bring the idea to the marketplace.

The product pitch Tetrick made was striking enough to garner the company $500,000 in seed money from Khosla Ventures, which is known in Silicon Valley for its early-stage investments in space, internet, and biotechnology companies. The firm was founded and is still run by Vinod Khosla, the billionaire co-founder of Sun Microsystems and former partner at storied venture capital firm Kleiner Perkins. Tetrick would later raise more money from Khosla, making it the start-up's largest investor and major influence over how the company would be run. He also persuaded big names including PayPal and Palantir cofounder Peter Thiel and Salesforce.com founder Marc Benioff to invest.

But Tetrick was also garnering interest from abroad. In Hong Kong, tycoons Li Ka-shing and Solina Chau were sprinkling promising upstarts with their considerable capital. The pair spent years overseeing a massive portfolio of industries, drawing money from transportation, real estate, financial services, retail, and energy and utilities businesses. Solina Chau is a sharp businesswoman with lucrative investments in an Asian mobile internet provider. Li Ka-shing is the enigmatic tycoon whose name has regularly featured at the top of Asia's richest businesspeople lists. He made much of his fortune in plastics manufacturing, real estate, and retail. The Hong Kong pair (Chau has described their relationship as being similar to that of Sancho Panza and Don Quixote) is reportedly worth more than $30 billion, and has a keen eye for disruption—they were some of the first investors in Slack and Facebook. Together, they run Horizons Ventures, a venture capital firm that had caught wind of JUST thanks to a man who, at the time, didn't work in the world of food at all.

The truth is, JUST might never have gotten off the ground had its founders not been lucky enough to bump into Sonny Vu, a Singapore-based, health-obsessed entrepreneur (he ingests up to forty supplements a day and is a self-described connoisseur of superfood powders) who made a name for himself in the wearable technologies business. Vu connected the young company with Horizons Ventures.

When Vu learned about JUST, he was blown away by the startup's ambitions to make a high-tech egg replacement. Chau was, too. "She actually flew out the next week to see them," says Vu.

After the introduction: radio silence. Vu said he didn't hear anything from Chau about how the meeting went, and JUST was mum for at least five weeks.

"And then, the next thing I know I see Chairman Li making scrambled eggs with Josh," he recalled. "I'd never seen anyone raise twenty million dollars in six weeks."

It's true. Balk and Tetrick had arranged a visit to Hong Kong. It was chaotic the moment the landing gear hit the ground. They stepped off their plane and realized they'd fully passed through the looking glass, stumbling into a world in which they say they could perceive a palpable and real interest in JUST's vegan mission—and in the hot new start-up funded by some of Asia's richest and most influential investors.

"It was like watching Justin Bieber get off a plane," says their good friend and fellow vegan activist, Paul Shapiro, a friend of both men who worked with Balk at the Humane Society.

You would not expect a pop star's reception for two middle-aged men peddling vegan mayonnaise and eggs.

"We were surrounded by people with cameras, iPhones going crazy, and microphones just being stuck in our faces," Balk recalls.

Reporters in Hong Kong were asking the same questions as reporters in the United States, curious about the fundamental environmental implications of these foods seeking to replace traditional meat and dairy products. How close were they to being a reality? And how soon would these new food technologies be in a position to threaten entrenched animal agriculture operations that had long commanded space in the food chain?

In time, so much buzz around JUST made it Silicon Valley's first and only food technology unicorn. Tetrick had found a direction, and for now he was committed to walking that walk. But not everyone was excited about him or his new company's mission to eagerly and loudly attempt to disrupt the animal agriculture system.

THE ART OF WAR

n the most Silicon Valley of stories, JUST was first headquartered in the garage of an unassuming little house at 371 Tenth Street in San Francisco—a short drive from the company's current headquarters. It was a shoestring operation back then. About thirty people were on the staff. A tiny food tech firm with an outsized goal, stuffed into a small nook in the city's SoMa district.

One day in 2014, a letter arrived. The garage was busy when it was delivered. It was from Unilever, one of the biggest food companies in the world.

The letter was polite, but clear. Unilever had noticed one of JUST's products, a vegan mayonnaise called Just Mayo. The food giant claimed that the name "Just Mayo" violated truth in advertising laws and standards of identity because the vegan product didn't include eggs. Its solution? JUST should change the name of its product. The "or else" was implied. The words ricocheted from staffer to staffer, stopping them cold and plunging the room into a confused silence. A shaky panic started to ripple through the garage, a reaction that was, in Tetrick's view, totally understandable, he says. Many of them had never worked in an environment where they'd experienced a letter of that type.

"They imagined the biggest, baddest bogeyman in the world," Tetrick said of his employees. The company panicked. "'Holy fuck they are going to shut us down. We can't sell it like this, no way. We must change it.'"

The problem with making eggless products that are meant to mimic and replace foods that typically would contain eggs is that you run the risk of pissing off the people who make the very products you're seeking to replace. And in the history of food innovation, established brands have rarely taken kindly to new entrants into their territories. That much was apparent in this situation, and Tetrick's small team felt the pressure.

So Tetrick was confronted with a difficult choice.

He decided to hold firm, responding to Unilever with a letter of his own that listed the reasons he felt his product did not need a name change. It was balanced on a razor-thin premise. He acknowledged that the government had a specific meaning for the term "mayonnaise," and that its definition included eggs as a fundamental ingredient. But JUST was selling "mayo," Tetrick said, a term that was not defined by the government.

Unmoved, Unilever began to ramp up ever more aggressive

language. The behemoth dashed off another letter addressed to Tetrick's garage, this time threatening a lawsuit, which it filed on Halloween in 2014.

Tetrick found himself at a crossroads—perhaps the most important he'd ever stare down. If he had given in, everything might have turned out differently. JUST might not even still be a company.

He turned to several people for advice, including some who told him that he should acknowledge that JUST was an insignificant plant-based food company compared to Unilever, and stood little chance of successfully fighting back against it. His products, at that point, were not widely available; in fact they were more likely to be found sitting on the shelves of obscure specialty food stores than in major, big-box retailers. Second, rather than tussle with an established food brand and bear the cost of a lawsuit, he should just cut his losses and call his product exactly what it was: a vegan dressing.

But Tetrick was allergic to the idea. Even as he mulled it, he says he couldn't stop thinking back to his early childhood. It became the basis for his argument that vegan mayo, as it were, had just as much right to sit on store shelves as real mayonnaise.

"This is where Alabama comes into it, right away!" he said. "No one wants to buy vegan mayonnaise in Alabama. Not a single soul."

An outside observer watching this drama unfold might come away impressed by a strident young leader committed to his principles and larger vision. And that perception wouldn't be entirely wrong. But ultimately, he didn't come to this decision on his own. He didn't mastermind the defense strategy. In fact, you can trace that to another person—and no surprise as to who that was. He picked up a phone and made a call to Washington, DC.

"Josh, what do we do about this?"

Tetrick might openly talk about how the nonprofit world was never a good fit for him, but the strategy he and Balk dreamed up together to combat Unilever was straight from the playbooks of nonprofit advocacy groups and activists. They suspected that, when consumers saw a massive outfit like Unilever nitpicking at a smaller start-up with a do-good message, they would be outraged. Tetrick figured that if he stood toe-to-toe with Unilever, he and his fledgling company—if they were lucky—might emerge stronger than before. In essence, JUST needed to mount the same tactics used by outfits like the Humane Society, where Balk worked, and PETA, only adapt them to a corporate setting. Campaigns use petitions and the media to their advantage, so why couldn't JUST? It was an effort to cast his company as the "little guy" in a story playing out on the American supper table. It was risky, of course. He and JUST could even win in the court of public opinion, but be buried in legal fees.

The company asked Andrew Zimmern, a restaurateur and food critic, to create, sign, and endorse a petition on Change.org. It was called "Stop Bullying Sustainable Food Companies," and eventually drew more than 111,000 signatures. "When a $60 billion company flexes its muscles to prevent a good-for-the-world startup company from succeeding, there are only two words for that: corporate bullying," the petition read.

A swell of interest in the petition drew the attention of the media, and before long the San Francisco–based CBS affiliate, KPIX, ran a short story on JUST and its battle against the larger forces within the world of food.

"For some fucking reason, the *Drudge Report* picks it up," Tetrick exclaimed, still stupefied by the event years later. The story on the *Drudge Report* linked directly to the CBS news story, which then went viral.

At the same time, food bloggers were homing in on the fight. Michele Simon, the future executive director of the Plant Based Foods Association, said she wrote and posted about the Unilever lawsuit about a month after it was filed. She sent a link to her post on EatDrinkPolitics.com to *The New York Times* food business reporter Stephanie Strom.

Just one day later, on November 10, the American newspaper of record published a story detailing to its wide audience the beef between Unilever and JUST. And a week after that, the Associated Press wrote a story of its own, followed by another story in *The New York Times*.

"I have *The New York Times* stories framed in my office," said Simon. "It was the most fun I'd had in my activist career."

As Tetrick remembers, for the next thirty or so days, people visited Unilever's official Facebook page to leave dozens of posts complaining about how the company was going after a much smaller one.

"I don't think they'd seen anything like this," Tetrick said. "We started this campaign, and thirty-three or thirty-four days later Unilever drops the lawsuit."

It was a stunning and important win for the people working at JUST.

The company figured out how to use its own nimbleness to try to outmaneuver an outsized threat, giving itself an advantage. And it successfully used then-emerging social media tools to build enough consumer interest in JUST's larger mission to mount an effective self-defense.

"How do you, in the art of war, use your enemy's weight against them?" Tetrick mused. "How do we shift this a little bit?"

Finding an effective strategy to achieve that goal, even if it wasn't totally his own idea, was paramount for Tetrick, who would soon find new battles to wage.

While Tetrick was occupied with Unilever, a government-regulated group that exists to promote the idea of eating more eggs was drawing up its own plan to attack the fledgling start-up. Unlike Unilever, though, the American Egg Board had its targets fixed on JUST for more than a year, and it was readying to deploy a sneakier strategy for handling Tetrick's ambitions.

The initial message was dispatched at 3:33 p.m. in the dead of summer.

Farmers had peppered her with complaints for some time. So after drafting the note, Joanne Ivy clicked send. And it was then, on August 20, 2013, that wheels were set into motion that would give Tetrick the entrepreneurial fight of his life.

"I'm getting a lot of emails about this product from egg producers and further processors," wrote Ivy,* who at the time was the president and CEO of the American Egg Board. "We think it would be a good idea if Edelman looked at this product as a crisis and major threat to the future of the egg product business and provide some advice and input as to how we should address this situation."

Edelman was the largest public relations firm on the planet by revenue. This is the same firm that provided crisis communications to Rupert Murdoch's News Corporation during its 2011 phone hacking scandal in the United Kingdom. The same firm that in the 2000s created a fake front group called "Working Families for Walmart" that was funded by the retailer to improve its image. The same firm that worked on behalf of Trans-Canada Corporation to launch campaigns in support of the Keystone XL pipeline.

Ivy was writing to other executive members of her staff specifically about JUST's eggless mayonnaise product, as well as rumors of the company's ambitions to one day sell an eggless scrambled egg product. The email sparked a two-year battle for the young vegan start-up, matching it against outsized forces that also included the USDA and the egg industry lobbying group. At one point, the American Egg Board conspired to pay

---

* Ivy declined to comment.

someone to try to convince Whole Foods Market to block sales of JUST's eggless mayo.

"Imagine the PR buzz that can be created if Whole Foods was on our side," wrote Elisa Maloberti, an Egg Board marketing executive, in another email.

And so the egg industry folks quietly designed and launched an all-out assault on JUST. In spite of the powerful players behind that effort, the strategy failed spectacularly. The victory against Unilever was a major win for Tetrick, but when the effort by the American Egg Board was eventually revealed, it only confirmed people's worst suspicions about tactics used by the established food industry to stymie creative new entrants into the space. The work of the American Egg Board was exposed thanks to a federal Freedom of Information Act request. The emails obtained through that request laid bare just how determined the $5.5 billion egg industry was at the time to thwart a vegan upstart. They paid bloggers to write disparagingly about JUST, and, at one point, egg industry executives were so upset they joked in an email, "Can we pool our money and put a hit on [Tetrick]."

"I thought people exaggerated about the crazy evilness of Big Food," Tetrick said. "Actually seeing it I was like, 'Wow. This is real.'"

But the effort to crack JUST never worked. Federal investigators eventually found that the American Egg Board and USDA had overstepped the boundaries set out by the US Congress for how a government-watched agricultural group should act.

Tetrick shepherded his company through these challenges as best he could, all the while luring more and more consumers to try his lines of plant-based condiments. For five years he grew his company and expanded the footprint of sales into mainstream retail outlets. In 2014, JUST's eggless mayo product launched in Walmart, Costco, Kroger, Dollar Tree, Safeway, and in select Hong Kong retailers.

Even still, it wouldn't be enough. Not for him.

8

THE LOST PUPPY

Five years after founding his company, Tetrick found himself standing in his kitchen, his palms pressed hard against a countertop. It was April 2016. He was seething, sad, and confused.

Just hours before, he had taken a solemn trip to Mission Dolores Park, accompanied by Balk, his brother Jordan, and friend Jill. They were all there for Jake, Tetrick's golden retriever of seven years. Riddled with cancer and near the end of his life, the dog's last moments would be spent at his favorite spot, a sunsoaked grassy knoll at the center of the city.

They let him wander for the last time in a landscape of lush green grass peppered by Mexican fan palms and Indian laurel fig trees. Then, as dusk settled quietly over the park, a veterinarian arrived.

Balk held Jake as the veterinarian administered a fatal shot of pentobarbital, which drifted through his circulatory system, allowing him a few last fleeting breaths before peacefully passing away.

Tetrick was in a haze. He sulked in his apartment at 1 South Park, just a few blocks west of the San Francisco Bay. In the next room over, Balk sat alone in a large, sparse living room with his own quiet thoughts.

"It was the first time I ever lost a loved one. Period. Like, in my entire life," Tetrick explained. "He was eight. For me, I was a lost puppy from the beginning. He was there when I was starting to become more the man I wanted to be."

He pushed hard against the countertop.

"What the fuck is going on with my life," Tetrick wondered.

From an outsider's perspective, there was a lot going on in Tetrick's life, mostly positive. His signature line of vegan condiments—including ranch dressing and mayonnaise—had already become household names, despite Unilever's and the egg lobby's best efforts. Product reviews were overwhelmingly positive.

In a blind taste test by the *Serious Eats* food blog, JUST's vegan mayonnaise was deemed more delicious than the real thing.

"Of the five mayonnaises we tested, it was the 'most balanced' with a good hint of acid from both vinegar and lemon to balance out its richness," the review said. "If all of [JUST's] upcoming products are as successful as this one, then they have a bright future ahead of them."

The Splendid Table was also kind to the company. In its review, the site said it could not tell the difference between JUST's plant-based mayonnaise and its usual go-to, Hellmann's.

It wasn't exactly the type of moment when one might expect a successful young CEO to experience a crisis. But Tetrick was grappling with the difficulties of growing a young company, along with deep-seated insecurities about his role and abilities. And a peek under the hood revealed a company dogged with allegations of scandal and uneasy with its leader.

About a year before Jake died, bad press coverage about JUST had started to pick up. In August 2015, *Business Insider* questioned the company's employment practices. And a few months after Jake's death, a series of *Bloomberg* stories published in 2016 alleged that the company misled its investors and operated a legally questionable product buy-back scheme to inflate the popularity of its products.

Stories like this have perpetuated whisperings within the Bay Area food technology scene. Very few of Tetrick's fellow food technology entrepreneurs are willing to go on record with their gripes, but it doesn't take much poking around to realize that Tetrick is not a beloved figure in some Silicon Valley food tech circles.

"I feel like he's lost credibility," Michele Simon told me in spring 2018. Simon once wrote blog posts ripping Unilever for filing a lawsuit against JUST. "Josh Tetrick is known for false promises."

Kurt Jetta, an analyst who tracks consumer packaged foods,

characterized Tetrick to me as a bullshitter, someone who cuts corners and acts unethically.

"That's my observation," Jetta told me in June 2017. "You can criticize the critics all you want. Eventually, if you don't act in an ethical manner, it's going to catch up with you."

So there he was, standing sullen in his kitchen just hours after putting his dog, Jake, to sleep. He asked the same question that's dogged him for years: "What the fuck is going on with my life?"

Grief threw his situation into sharp relief. Could he do more? Could he be more? And again Tetrick turned to Balk, who sat quietly in his living room, thinking over the events of the afternoon. The two men began to brainstorm. They burned through a few ideas, one of which included developing their own plant-based meat alternative, something akin to the Beyond Meat burger or Impossible Foods products. But, in the end, they decided ardent meat-eating people would ultimately want more than just plant-based imitators.

And that's when Balk posed a question: Why not explore the possibility of getting into cell-cultured meat?

Tetrick was immediately receptive to it. The idea inspired him.

"It was about creating an economy that's not built on the backs of animals," he says. "Make capitalism work for animals."

They'd already worked on the model that used plants to replace the need for using certain egg products and found success. But could they effectively expand into animal cells to grow sustainable animal-based products, too, and more radically?

Shortly after his conversation with Balk, he called up New

Harvest, a nonprofit group dedicated to advancing the field of cellular agriculture, to speak with a few tissue engineers. Those conversations ultimately led to hiring Eitan Fischer, whose initial job was to figure out whether it would be feasible for JUST to create and fund a laboratory dedicated to this work. Fischer came back to report in the affirmative, and subsequently took on a leading role in establishing JUST's cell-cultured meat laboratory, which initially operated in stealth mode at the company. Only nine people at JUST knew.

Tetrick decided to keep it secret because it felt like such an unknown for the company. There was, of course, the chance that within six months the project would be a failure. But if it succeeded, he wanted it to feel like a structured department within the company before it was announced to the entire staff.

The secret lab operated out of what was previously a large storage closet that had been outfitted with a freezer system for storing cell lines, miniature bioreactors, and stations for scientists to examine how cells grow under different circumstances and in different liquid mediums. It's still the space being used today, in addition to a new lab built on the first floor just below it, only now it looks more official. A large photo of Jake hangs just inside the doorway on the wall.

Tetrick's bouts of inspiration haven't been enough to quiet occasional naysayers, who have been able to amplify their voices through Silicon Valley's media, particularly in stories published by *Bloomberg*, where disgruntled former employees have happily sniped at their former boss. In spite of these attacks, generally anonymous, no one has been able to prove that Tetrick is guilty of cutting corners or of any wrongdoing.

Within his own company, Tetrick's management style had been called into question, too. To be clear, Tetrick admits that he's not a natural-born leader. Not much about how Tetrick describes himself makes him seem like an easy person for whom to work.

"I don't know if it's because of how I was raised, I don't know what the fucking reason is, but if things are too calm, I feel very bored," Tetrick said.

"If we're not running on the edge then we're never going to do anything, but if we run too close to the edge every single day, people cannot mentally and physically handle it. I think that's the challenge: How do you get people close enough all the time? I find that balance difficult."

But the edge is a difficult place for people to spend their workday. And there's been a lot of turnover at JUST in the last several years. In the two years it took to report this book, the cell-cultured meat laboratory alone experienced 100 percent turnover, with two of its leading scientists leaving the company to launch their own cell-cultured meat start-up. (The company defends itself by arguing its cell-cultured lab is now staffed with more people than before.) Online employee feedback resources, such as Glassdoor, feature numerous (anonymous) complaints about Tetrick and his management style.

"The trends at [JUST] should not be normalized, nor should they be compared to successful companies with intelligent and proven leadership," one former employee wrote. "What's happening at [JUST] is a symptom of a very serious disease. And I think we all know what that disease is."

The scathing review continues: "We quit because Josh Tetrick is a lying, manipulative, thin-skinned, attention seeking, reactionary, incompetent, Steve Jobs wannabe who reminds me more and more of a certain 'tweeting President' each day, and we all reach a point where we can no longer manage the cognitive dissonance of trying to be a decent person while also enabling a con-man on a daily basis."

Disgruntled employee issues have also veered into the personal. In 2013, company drama went public when an employee named Javier Colón* was fired. According to a report by *Bloomberg*, the dispute between the two was over employment contracts and how Josh attempted to change the wording of them to give employees three weeks of severance instead of three months in the event they were let go. Colón filed an unfair labor action against JUST which was settled privately.

Colón was also allegedly roped into personal drama with Tetrick, as he was friends with his boss's girlfriend at the time.

According to that same *Bloomberg* report, when Colón's company phone broke, Tetrick told him to use another company phone—only the one Colón picked still contained text messages revealing Tetrick had been having an affair with one of his employees. Colón informed his friend who was dating Tetrick.

---

* Colón did not respond to a request for comment.

Angry, she demanded Tetrick fire the woman he'd slept with, something he refused to do, texting that "Khosla would hang me—it is a huge lawsuit."

Khosla is, of course, Vinod Khosla, the prolific Silicon Valley venture capitalist and major early investor in JUST. In some ways, Tetrick's entrepreneurial original sin might have been courting Khosla. It enabled growth, but the growing pains were difficult.

When Tetrick first approached the firm, he knew very little about how to run a proper business, to say nothing of how to navigate a landscape of Silicon Valley investment titans. He says he had less than $3,000 in his bank account and a burning desire to make his and Balk's idea work. From Tetrick's perspective, it put Khosla in a terrific negotiating position.

He signed a set of documents that laid out the terms and conditions of receiving funding. In exchange for Khosla's support, Tetrick would sit on his company's board but would relinquish 40 percent of the ownership of JUST, as well as a significant amount of control, as well as representation on JUST's corporate board.

"They could hire, they could fire, all of the key things," Tetrick said.

And he signed on the dotted line, he says.

From the first dollar Tetrick raised for JUST in 2011, all the way through the company's Series C funding round in December 2014, he says Khosla retained the ability to veto ideas that would make fundamental changes to the company.

"You don't realize that you're actually stitching together your own noose sometimes," he said.

Ultimately, Tetrick found that the power Khosla Ventures wielded was the authority to issue a two-letter word that he was forced to obey: "No." As Tetrick puts it, if he wanted to take the company in a different direction, he was required to seek permission. If he felt it would be wise to jump into the world of cell-cultured meat, for example, he couldn't unless Khosla Ventures and other big investors gave the stamp of approval. And that irked him.

When I asked Khosla Ventures about this characterization, the firm demurred.

"It was never a control thing," said Samir Kaul, a cofounder of the firm and former member of JUST's board. "Josh had control."

Kaul said he did have concerns about some of Tetrick's business strategies. When JUST first started, for instance, Kaul said Tetrick's goal was to make a nonanimal ingredient to replace eggs in a variety of products, including scrambled eggs, that could be sold to college campuses across the United States.

"We were at a conference together and we went for a walk," Kaul recalled. "And I said, 'Josh, if you're not going to go directly to the consumer, we're not going to fund you anymore.'"

He said Tetrick altered course and not long after was selling his initial lines of vegan condiments. The move was seen as a success. Khosla recognized this as a solid and safe business. Investors considered JUST's future bright, knowing that they could make a lot of money if JUST was bought by a larger corporation. But Tetrick still wanted to push boundaries.

Uninterested simply in cashing out and making a lot of money, he pushed to continue growing the company. The idea was that you can't disrupt the status quo of animal agriculture if you don't keep taking swings. Beneath the thumb of venture capital, he said he felt it was tougher to run a nimble business. He wanted to achieve the original mission of the company, and to do that the power structure of the company needed to be shuffled through the arcane vagaries of corporate governance. That opportunity presented itself when JUST entered its Series D round of funding in August 2016.

Tetrick said he began talking with Nan Fung, an investment firm comprised of a syndicate of companies that is headquartered in Hong Kong. The firm—along with China Construction Bank, New World Group, and others—was willing to pour a lot of capital into the company, enough to multiply JUST's valuation from $190 million to more than $750 million.

As Tetrick explains it, the deal he drew up with Nan Fung reshaped the corporate board, increasing the number of company-controlled common seats on the board to outnumber investor-held seats. The way it worked out, JUST would have more influence over its own future than its investors. And even though Khosla Ventures would lose some percentage of its total control in the deal, the idea was that it would, all in all, be good news for everyone involved because of the massive influx of capital.

Khosla Ventures had a big decision to make, Tetrick said. It had invested in the company when it was only worth around $3 million; the agreement then gave it about 40 percent control. Now, many years and products later, the lucrative deal led by Nan Fung presented Khosla with an interesting proposition: It could vote to approve the financing (which meant also approving changes to the board's structure, diminishing its sway over the company), or it could vote the idea down. Approval of the deal required a simple majority of the board. Tetrick said he believes he could have achieved it without Khosla's vote, but having their support would make the approval smoother.

Khosla decided to approve the deal.

This was a monumental moment for JUST, one that would ultimately allow Tetrick to more easily alter the company's direction. In just a few years, JUST had risen in prominence from a garage-bound vegan mayonnaise producer to become a fully operational food company.

"It was that moment where it moved from being a company controlled by venture capitalists to—and I'll say what it is—a company controlled by activists. Truly," Tetrick says.

The following year, in spring 2017, the company closed a Series E round of funding, which put it at a valuation of about $1.1 billion, making it Silicon Valley's only food technology unicorn.

"I certainly wouldn't bet against him," Kaul quipped about Tetrick, adding that he maintains faith in and respect for the young CEO—that he's capable, passionate, and mission driven.

Tetrick deftly navigated those early hurdles, though not without a cost. Impressive funding rounds and a bigger valuation meant extra attention. The company was the subject of multiple stories—most of them printed, again, by *Bloomberg*—that cast Tetrick in a negative light. One story even had an online interactive feature in which the reader could throw virtual mayonnaise at his face.

That story was based on an anonymous rumor that the company had engaged in an illegal product buy-back scheme. It alleged that Tetrick enlisted and paid people to buy his products to inflate the sense of consumer interest, and it had very serious consequences. Curious about the stories it was seeing, the US Department of Justice and the US Securities and Exchange Commission launched their own independent investigations into the company, both of which closed their work in March 2017, concluding the allegations were immaterial.

This new working knowledge of navigating Silicon Valley—and the press covering the tech ecosystem—made Tetrick a savvier businessman, someone more attuned to the intricate contours of the legalese that defines a young tech start-up. And this brand of corporate street smarts would come in handy very quickly, because shortly after Khosla lost its sway over the company, Tetrick found himself staring down a coup d'état* from within at around the same time the company completed its Series E funding. It was led by Jim Flatt, then the chief technology officer; Lee Chae, the then–vice president of research and development; and Sofia Elizondo, the then–vice president of business development.

---

* Flatt, Chae, and Elizondo did not respond to a request for comment.

"For whatever fucking reason they decided to work with at least one member of the board and other investors around us to actually change the corporate governance documents back to Series C—not the valuation of the company, but the control of the board."

Put simply, three senior staffers did not like the direction in which Tetrick was steering the company and, according to Tetrick, operated behind his back to remove him from power. Had they found success, it is possible that JUST would no longer be controlled by Tetrick, would possibly have been acquired by another company, and instead of marketing its own products might have relied more on licensing JUST's technology out instead. The board member representing Khosla Ventures was not the person the three employees turned to, Tetrick later clarified to me.

"I don't think they believed that the best way for us to be a company was for us to sell our own products," Tetrick said. "And they certainly weren't down with the cell-cultured meat idea."

To their credit, Tetrick says they made a good argument. In some ways, licensing JUST's business to be manufactured and sold under a different name might have been a better route to go than branding, marketing, and selling food products on their own. In doing that the company could have saved a lot of money. Tetrick said he might have been able to dismiss members from among the operations, finance, research and development, and manufacturing staffs. There's a lot of money that goes into making a physical product, and going down the licensing route might have helped the company get some of its overhead down.

Instead of burning $3 million they could have burned about $1 million.

"What was irrational was how they went about it," he said, describing how they connected with investors and members of the board to create a fait accompli.

The three employees seem not to have anticipated that they might be caught by Tetrick before their plan hatched. Because information—some accurate and some inaccurate—seemed to be leaking from the company and to the press, Tetrick hired a security firm called GRA at the recommendation of the former head of security at Apple. The firm was tasked to do a full investigation. What turned up was something completely different: emails exchanged by the three employees that revealed the plan to overthrow Tetrick.

They were having dinner and drinks with JUST's investors and members of the board, trying to sway their opinion.

An email from one of the three employees dated March 2017 was penned to a family member. "One reason I'm working so hard is there is a chance for really big payoff."

In emails a month later between two of the employees, they joked with each other about the future of JUST:

"When governance changes you may have to take over as chief operating officer," one wrote.

"And CFO. And CMO."

"And facilities? Clean-up aisle 3 . . . oh and 4 . . . and 5."

"Chief of janitorial services."

Tetrick was livid. He fired the three employees and asked one member of the board to step down. Ultimately the entire board was reshuffled, with many of the members—including former Obama administration official Kathleen Sebelius—opting to stay on as advisers to the company rather than full board members. The board was eventually reconstituted, but Tetrick is now in control of the company's fate in a way he wasn't before.

And as for Platt, Elizondo, and Chae, the three of them left JUST and started their own company, Brightseed, which uses machine learning to dive deep into the genetic makeup of plants to figure out which compounds can be used to meet certain needs in new products—not too dissimilar from JUST's Orchard system.

When he thinks back to the coup, the government investigations, the rumors, and the bad press, Tetrick tenses up.

He's seated on a sofa in the middle of JUST's busy headquarters. It's well past six p.m. and the room is abuzz with activity. People at their desks working, chefs maneuvering around the corporate test kitchen, someone walking into the cell-cultured meat laboratory. Dogs were still running around the office, too.

"It's a shame you need that crazy thing to make you more thoughtful and more mature, but I needed it. I really did. I needed to go through something crazy." The team seems to have stabilized, too, or at least coalesced around Tetrick's vision. Tetrick says, "I feel really proud. I can look around at these people who are here now and I can have more trust."

The milestone event ushered JUST into its next phase, making cell-cultured meat the priority. But is he up to the challenge?

Tetrick, who hopes and pushes so hard to be the first to introduce this new meat to kitchen tables, is not the obvious pick to be the poster boy for it. He isn't a Steve Jobs–style genius. He's not a master of supply chains like Jeff Bezos. And he doesn't have the vision of a Bill Gates. And even within his own sphere, he lacks the scientific know-how of Mark Post and many others.

Is he capable? Yes. Tenacious? Sure. But the natural leader of this new movement? No.

This is the curious thing about Tetrick, who founded and runs one of Silicon Valley's most successful food technology start-ups, but whose expertise has never propelled it.

But that hasn't stopped him from pushing forward. When he feels he has purpose, Tetrick is a go-getter. He's gotten to where he is because he's kept asking questions of life, some inner agency that has pushed him to search for new ways to carry out his personal mission. That desire and drive can't be ignored. Neither can his capacity to tell a great story, which has remained one of the strong suits that differentiates him from the CEOs of other cell-cultured meat companies.

The catastrophe in the Netherlands showed that the status quo would be difficult to overcome. But Tetrick still aimed to get cell-cultured meat to the market first, hopscotching all over the world to sweet-talk regulators into allowing him to sell and serve his meat.

But there are many other companies, and many brilliant scientists and entrepreneurs racing him. These start-ups have perfected their own versions of cell-cultured meat, and they've also been eyeing the regulatory landscape.

I started to schedule trips to other Bay Area start-ups, to Israel, and to Europe. All the while, Tetrick stayed focused on finding a way to achieve his goal: getting cell-cultured meat to market first.

9

THE REST OF
THE HERD

Somewhere over the Aegean Sea a soft-spoken flight attendant leans over to ask:

"Pasta and beef, or couscous and chicken?"

I choose the latter. He hands over a tray with pita, hummus, and a plastic bowl of salad featuring carrot shreds clinging desperately to roughly diced cucumber. A cookie and small cup of water play supporting roles, and a warm, sealed aluminum tin contains the main course.

I start at the edges, pretending I'm still at a café in Tel Aviv along the leafy Sderot Ben Gurion thoroughfare. I quickly migrate to the hummus before peeling back the lid on the tin. It contains precisely two colors: beige couscous and red tomato sauce covering three chicken meatballs. I slice through a meatball and find exactly what I'd expect: white processed chicken meat.

A few days earlier I had met with Didier Toubia, the cofounder of Aleph Farms, a cell-cultured meat start-up with a laboratory in Rehovot, a small city a little more than twelve miles south of Tel Aviv. "Meat paste" he had called it then, the stuff most of the start-ups are making; muscle tissue cells in a large enough mass to make a meatball. If cell-cultured meat fails in the main market, perhaps they can sell it to the airlines, I mused, poking another meatball and watching it give way. Hopefully it wouldn't come to that, Willem van Eelen's dream finally realized, drowned in red sauce somewhere in the sky.

But after a week and a half in the Holy Land, it had become abundantly clear that a cohort of Israeli start-ups pose a major threat to the Silicon Valley companies chasing the cultured meat dream. If it wasn't an American company getting cell-cultured meat into the market first, it'd most likely be a company in Israel. Regarded by many as the world's second Silicon Valley, Israel is home to three cell-cultured meat start-ups: Aleph Farms, SuperMeat, and Future Meat Technologies (there are eight in the United States that I've been watching closely).

I meet Toubia in the middle of a warm June afternoon. We're on the third floor of a nondescript building in an office park where, at the time, Aleph Farms shared laboratory space with another company. On a separate floor in the same building, another cell-cultured meat firm, SuperMeat, also operates.

The Israeli government is proactive about giving bright and potentially disruptive technological ideas financial assistance at their earliest stages. Looking to stimulate the development of the technology, the Israeli government has provided numerous grants, as well as offered support through its financing of a $25 million food technology incubator, the Israeli Innovation Authority. Each year, the Innovation Authority gets about half a billion dollars to dispense as conditional loans to hungry and ambitious start-ups, large or small, across a diverse array of sectors. If they fail, the government doesn't expect a return on any of the money. If they succeed, they pay the government back.

Anya Eldan manages the start-up division of the authority, and she has a nose for disruption. "If we got back 100 percent of the money, we'd think we're not taking enough risk," she told me during one of my visits. "Our role is to build a successful, competitive innovation industry."

There is no comparable American program.

Aside from the obvious economic upsides of cell-cultured meat technology taking off, Israel itself recognizes that its success would bolster the nation's own food security. Limited land and water resources stop it from being able to produce most of its own food. Every year the nation posts a sizable trade deficit in food goods. The situation is unlikely to improve if the country doesn't take action.

At the end of 2017, the Global Food Security Index published by *The Economist* Intelligence Unit and DuPont found that, while food security has generally improved over the past decade, the future poses risks for countries in the Middle East, North Africa, and South America. Unpredictable global economic growth, increases in inequality, and forced migration due to climate change and political turmoil are expected to impact food security negatively across many regions of the world, some worse than others. Nor is there enough food to go around. The United Nations predicts that the planet will need to increase its agricultural production by about 70 percent in order to feed the world's people by 2050.

Israeli imports could change if the cell-cultured meat companies there are, in the future, able to grow the same meats within the national borders. So while US companies are often driven by vegan ambitions, the Israeli companies are largely motivated by long-term self-preservation.

Aleph Farms itself was actually started in the north of Israel, in Haifa, by a prominent researcher named Shulamit Levenberg, who is also the dean of the faculty for the biomedical engineering program at Israel's elite Technion Institute of Technology. Her expertise in understanding and constructing complex vascular systems—the arteries and veins that reach throughout the body—has injected Aleph Farms with an intellectual rigor and ambition that is unique among its competitors. While JUST and Memphis Meats focus much of their attention on what Toubia, Levenberg's cofounder, would call "meat paste," for processed products such as chicken fingers, duck carnitas, and nuggets, Aleph Farms wants to skip growing cells in a liquid suspension system altogether. Rather than grow clumps in cylindrical bioreactors, Toubia explained that the main goal of his company is to grow beef as a 3-D structure—also known as steak.

JUST and many of its American competitors grow meat by

putting cells into a nutrient-dense liquid where they sit and pro-liferate, but Toubia and Levenberg's ambitions require a more complex system that closely mimics what happens inside a living animal: a man-made blood-vessel-like vascular system that pushes liquid medium to the cells, providing nutrients. The benefit of set-ting ambitions so high was a no-brainer for the start-up. It might take them a little longer to release a product into the market, but it will position them to be perhaps the best at making a more complex cut of meat. Toubia has said he expects a full-scale global launch of his company's steak products sometime in 2023, though he's still unclear on which market to enter first.

In the business, the vascular system needed to grow meat is often referred to as "scaffolding." Toubia likens it to growing food via hydroponics, which is a method of growing plants with-out soil by using nutrients mixed into a water solvent. It might sound strange, creating such an unconventional system, but Aleph Farms doesn't shy away from novelty. As Toubia puts it, technology has always been used to try to improve the natural process. The question people should be asking, he posits, is not whether a food is natural or not. Rather, they should be asking themselves, "Is this food good or not?"

There are, of course, complications in opting to use the vas-cular system. "We've had to start almost from scratch," Toubia explains, as mammalian cell lines are far more persnickety to manage than cell lines from birds. The companies that are mak-ing chicken focus mostly on growing muscle cells, but creating steak means that Aleph must also perfect growing fat cells.

It's this spirit of inventiveness that distinguishes the work happening in Israel from the rest of the world. The Israelis are truly pushing the boundaries of science. They're making a bet: They won't be first, but they hope they will someday be the best.

I woke up a little earlier than usual on my final day in Israel. The sun rose beyond the scattered towers of Tel Aviv, unleashing a wave of pink to wash westward over the city and eventually to the edge of the Mediterranean. The boardwalk bustles with early-morning types, tackling their usual exercise regimens. Sculpted men sprint along the coast, and pockets of classes practice their yoga and stretching. I stroll to a nearby coffee shop where I order a typical, vegetable-heavy Israeli breakfast. I'm here to meet with Rom Kshuk, who leads one of the cell-cultured meat companies that is near the head of the pack racing to get a product to market first.

Kshuk lives just outside Tel Aviv, but his company, Future Meat Technologies, is based in Jerusalem. Yaakov "Koby" Nahmias, a biomedical engineer and professor at Hebrew University in Jerusalem, founded the start-up in 2017. Instead of focusing on growing muscle cells, Future Meat Technologies is leading the way on the development of technology to grow connective tissue—which needs fewer ingredients and less liquid medium.

"We would like to sell farmers a machine one time," he says. "I would sell them capsules with different cell starters, and they could use it to grow meat."

Future Meat Technologies seeks to be a business-to-business company, selling knowledge and machinery to other companies seeking to make meat and market it under their own brand. The first products would be—like the ones JUST is working on—mincemeat. Once that is perfected, the company would focus on full cuts. The decision to go this route was an easy one, said Kshuk. Why go up against the likes of Tyson on the retail front when you can impact the global food system from the inside?

In October 2019, the company got $14 million in Series A funding to begin doing just that—the second largest Series A funding round of all the cell-cultured meat start-ups (Memphis Meats raised $17 million in 2017). Kshuk said much of that money would go toward building the world's first cell-cultured meat manufacturing facility in Tel Aviv.

The funding round was led by Chicago-based venture capital firm S2G Ventures, which was joined by Chinese food and agriculture venture capital firm Bits x Bites and Tyson Foods' own forward-looking firm, Tyson Ventures.

"This continues our investment in Future Meat Technologies, which is focused on disruptive technologies related to our core business," Tyson Ventures representative Amy Tu said at the time. "We are broadening our exposure to alternative ways of producing protein to feed a growing world population."

With a production facility built and running sometime in 2020, Future Meat Technologies says it hopes to have an early product ready by the end of 2020, with more available in 2021. Unless Tetrick beats them to market, that puts them at the forefront of the industry, and, if those early products are first sold and eaten in the company's domestic market, it could make Israel a top contender for winning the edible space race. That has its pros and cons. The upside is cell-cultured meat would get to consumers in a tech-friendly market, which may help persuade regulators elsewhere to follow suit. The downside, and one of the reasons companies in the United States aren't doubling down on getting into the Israeli market first, is that it's extremely small and doesn't represent a natural launching point into more heavily populated regions, such as the European Union or China.

Regardless of all the impressive technological work happening around the world, there are still a lot of unknowns—as Mark Post made clear to me. After introducing the world to the concept in 2013, Post started his own company, Mosa Meat. It's headquartered about 130 miles south of Amsterdam in the midsize city of Maastricht, where the treaty that formed the European Union was ratified in February 1992.

I struck out for Maastricht in the quiet hours of a chilly morning, clumsily navigating Dutch signs in Amsterdam's Central Station to find a sleek yellow and blue railcar. I sank into a cushy gray seat.

The train pulled from the station and chugged through the city's sleepy suburbs, the electricity of Amsterdam's busy city center giving way as we sliced through the flat Dutch countryside at a swift clip. Bucolic green fields lay blanketed in the pleasant glow of a sun stretching its sharp pinks and oranges outward. Within an hour the sky would settle on a crisp blue. The train passed a small herd of grazing cows and I closed my eyes and listened to the sounds of commuters stepping on and off the train, the unfurling of their daily newspapers, and muted conversation.

By the time I woke up we were about ten minutes away from my destination, enough time to wipe away the remaining sleep and figure out just how I'd be getting to my meeting on time.

Maastricht is a cozy town of about 280,000 people situated on the Maas River, right along the border with Belgium. It claims to be the oldest city in Holland (something disputed by residents of Nijmegen, two and a half hours north), and its cobblestone streets lined by weathered homes look the part.

The morning is clear and cold. Pulling my jacket tight, I push upwind toward the doors of an imposing academic building that houses Maastricht University's physiology department. It's a hulking, modern structure that stands out on an otherwise flat landscape. An odd stairway leads up to the building's front entrance that has me feeling as though I'm about to board a UFO rather than enter a prestigious academic institution.

Unlike most cell-cultured meat companies in the world, Post's is part owned by the public university—which is to say it tends to move a lot more slowly compared with the more nimble cultured meat start-ups that aren't tethered to academia. The arrangement is akin to having a licensing agreement, in which Dutch taxpayers get something back for their investment of public dollars into his research.

Post himself is a towering figure, tall as a doorway. He peers at me with his gray-blue eyes from behind a pair of modern rimmed glasses. His voice is deep and deliberate—fitting for a man recognized the world over as one of the pioneers of cell-cultured meat. After all, he worked directly with Willem van Eelen, and it was Post who in 2013 publicly unveiled the world's first cell-cultured meat, a proof of the concepts van Eelen had promoted his entire career. I showed up at his office curious to learn what he's been up to since that high-profile unveiling.

The answer is fat.

That first patty he created was comprised entirely of muscle tissue. But it's the fat in our meat that gives the food much of its texture. And about half the flavors in meat come from the aromaticity of fat cells, not muscle cells. Without fat, cell-cultured meats from beef, chicken, pork, and other animals taste pretty much the same. Once you add fat, though, it's much easier for people to differentiate the species by taste. The problem is that growing fat is different from growing muscle, he explains.

"There's not as much knowledge about growing fat tissue as there is muscle tissue," he says.

His advancements in the field drew attention from important American and European industry players. In 2018, Post secured $8.8 million in investments to bring cell-cultured beef to the consumer market by 2021. Much of that money came from the pharmaceutical company Merck, as well as from Bell Food Group, a Swiss meat processor that has operations peppered across the continent. It was Merck's first investment in the food industry, and a signal that it sees a potential future in cell-cultured meat technology by providing Mosa Meat, and perhaps in the future other companies, with effective growth mediums.

"Meat demand is soaring, and in the future won't be met by livestock agriculture alone," said Lorenz Wyss, the CEO of Bell Food Group, in a statement announcing its investment. "We believe this technology can become a true alternative for environment-conscious consumers and we are delighted to bring our know-how and expertise of the meat business into this strategic partnership."

One of the other key aspects distinguishing Mosa Meat from the other cell-cultured meat companies is that it's run by meat eaters. Mark Post is not a vegan. He enjoys eating steak, plain and simple. His primary reason for working in this field is concern for the future of food security and the environment. His ultimate goal—which he's unabashed about, even in the presence of cattle ranchers—is to reduce the global cattle population from half a billion cows to about thirty thousand, some of which would be used for cell collection for the production of meat.

Most of the American cell-cultured meat companies have been so sensitive about their outward messaging and marketing that the executives who run them too often speak in rosy terms about their work. If they've run into any technical difficulties or scientific quandaries, they aren't necessarily forthcoming about it, even if I ask head-on. There's a sort of veil that exists between these companies' public-facing optimism about this new product and the reality that what's happening under the microscope is actually extremely complex and prone to all sorts of hiccups big and small. Post is uncharacteristically blunt, though, perhaps because he is an academic at heart. He revels in thinking and talking about cell-cultured meats, acknowledging the challenges and intellectual puzzles to solve before it will be ready to be served up sizzling to the public.

He earned his medical degree from Utrecht University in 1982—specializing in blood vessels—with every intention of becoming a surgeon. But after a couple years in patient care, he decided that it wasn't the path for him. Being a surgeon might have promised social esteem among his peers and a high income, but as Post put it, having patients is nice but their problems are boring.

"They intellectually are just not challenging, and I'm too egotistical to deal with them," he tells me. "The problem solving and level of pragmatism eventually doesn't really work for me."

I can't help but chuckle at this refreshingly straightforward answer.

Instead of working within the confines of an operating theater, speaking to fellow doctors, Post has spent the last few years debating with and learning from politicians, cattle ranchers, marketers, sociologists, meat industry executives, investors, and many more people he otherwise wouldn't have met.

Across the hall from his office is his laboratory. Post swipes a security pass to unlock a heavy door that leads into a room that looks familiar; every cell-cultured meat lab looks basically the same. There are microscopes, there are stainless steel cell-culturing stations, little vials and petri dishes filled with liquids and growing cells, and assistants wearing white lab coats. Of course, it's the end result that truly differentiates one company from the next. Post's goal is to grow beef, without the fillers common in high-tech plant-based burgers (like Impossible Foods) that compensate for hard-to-grow fat. Beef sets up the challenge of figuring out how to make an iconic food item: the burger.

"I think creating a pure product helps with transparency," he tells me. "If you start to make hybrid products it becomes a little bit murkier for consumers. We don't want to disturb our message, or contaminate it with a 'Oh, by the way, this also has, you know, 20 percent coconut oil' or something."

Eager to have a little fun with my own curiosity, I ask Post what sorts of questions he never gets from inquiring minds.

"Nobody has ever asked me, 'Is this cloning or not?'" he says. "Nobody has."

I take the bait.

"Is it?"

"Of course it is," he says, causing us both to laugh, albeit a bit nervously.

"From a biological point of view, cloning is nonsexual reproduction," he explains. "Well, this is that. I'm happy they don't ask that question because then I have to explain, and in the minds of people it means something else than when it's in my mind."

"Is there anything inside an animal that happens that humans just can't replicate?" I ask.

At this point he's seated in a chair. He leans back and looks to the ceiling for a moment.

"Yeah, there is," he replies. "Any entire organism is incredibly complex, and we don't really understand every detail of their complexity."

So sure, this makes sense to me. We understand some of the mechanics of what happens within the anatomy of a cow; we have insights into how certain bodily chemicals interact with other bodily chemicals to make life tick. Many more aspects of that anatomical universe, things playing out inside pretty much every living thing, aren't totally clear to us. We don't even know what we don't know. For instance, how important movement might be to animal tissue. A living animal moves around as it grazes and wanders. Is there something inherently special about that basic act of living? Does it affect the protein composition within the physical tissue of the animal, and in a way that defines its meat and what it is on the most minute, molecular level?

I recall a conversation I had with Bay Area–based Finless Foods CEO Mike Selden, one of the most thoughtful leaders in the industry. He told me that his company's scientists had done a lot of work researching how an animal's movement is incredibly important to the quality of its meat. It's one of the reasons why we don't eat farmed tuna, he explained. An adult bluefin tuna can grow to about ten feet long in the wild, where it swims against strong currents for thousands of miles. When you stick that creature in the small confines of a farm, physiologically it develops differently, which affects the quality of the meat.

As Post explains, there are also still questions around the lack of an animal's immune system. "There is no immune system in this cell culture system," he says. "Is that important? We don't know."

But the most intriguing question that he poses has to do with some of the most elemental aspects of cell culturing tissues, which is that without that incredibly complex organism around the cells, there are certain roadblocks humans are just going to have to learn how to grapple with and solve.

"I see one hazard, which is that we rely on the replication of cells a lot," Post explains. "With every replication, that DNA is kind of sensitive to mutation. We know that from different examples, that if you rely on replication a lot then you are going to have genetically altered cells, or unstable cells."

This could develop into an acute problem for the makers of cell-cultured meats. Scientists in these laboratories have noticed that some cells from certain species take on genetic differences after a few replications. In general, a full cell cycle can take several hours to run its course, but once you have a number of cells replicating at the same time, the growth is exponential. In other words, the easy theory is that cells replicate and create carbon copies of themselves—but the reality is that they don't actually act that way. In mice, for instance, that mutation happens quickly, after about four generations of cell replication.

Naturally occurring DNA replication is an incredible phenomenon in life. Every time a human cell divides, it copies and transmits an exact sequence of about three billion nucleotides to the newly formed cells. Every now and again, though, life will throw a curveball—something happens and polymerase enzymes will insert too few or the wrong nucleotides into the new DNA sequence. Most of those editing errors are fixed naturally because DNA has its own built-in repair shop for just this sort of situation. But sometimes these mistakes make it past the repair stage. For years scientists have worked to figure out how and why.

"Fortunately, we have not seen that yet from the muscle cells from cows, pigs, or humans for that matter," Post tells me. "In human cells and large mammals, it doesn't seem to happen. This is not only my observation; other people have made exactly the same observation. But you kind of have to assume that it's going to happen, so you need to deal with that. You need to ask that question."

He stops to assure me that eating a genetically altered cell is not going to be a threat to my health. After all, the meat and plant cells that we commonly eat today are dead once the plant is picked or the cow is killed. So as he explained, it's not a concern that a person would eat a living mutated cell and then develop a health problem because of it. Mutations present problems not because they impact health or taste, but because they make scaling production much more complicated.

There is some question around whether such a mutation would wind up impacting the nutritional value of any resulting cell-cultured tissue, but these are not problems that he's currently observing in his own laboratory while working with his cultures, or in the prototypes that he's been developing for tasting.

"But the fact that I cannot see [these potential problems] does not mean that they are not there," he says, ever cautious.

I recalled a moment in Post's office when I'd asked him about Tetrick, and about whether he felt comfortable that someone had promised so publicly to get cell-cultured meat to the consumer market first. His reaction was easy to understand; it was the same reaction everyone in this nascent industry had to my question. It all depended on the quality of the meat Tetrick would put out, because whatever that product was going to be, its quality would be highly scrutinized.

"That's my biggest concern, I can't control it," Post told me. "What I have heard about Josh is that he is pretty much uncontrollable. . . . I think you have to be really careful about this."

And how could Post not have that impression? I wondered to myself. Any studious reader of news coming from Silicon Valley would have every reason to feel skepticism about the operation Tetrick was running, with its turnover, investigations, and messy attempted coup d'état.

The truth is that, for as much as the entrepreneurs in the cell-cultured meat space swap gossip about one another, nobody truly knows what's happening inside their competitors' labs. Reality is more nuanced than rumor. Post's skepticism about Tetrick isn't difficult to understand. JUST's problems have been aired in the media, the scientists who founded the company's cell-cultured meat laboratory have all left to start their own company, and Josh's competitive spirit around getting his meat to market first only increases the pressure felt by the rest of his peers.

But Post's instincts are correct in that the companies must be mindful about exactly which products are released first. Truly, the entire industry needs to be careful—not just with the timing and quality of their work, but also how they market it. There's clearly skepticism and fear of cell-cultured meats among ranchers—many of whom had made rearing cattle a family business—and meat scientists. Even as the leaders of the cultured meat start-ups began to signal that their products were nearing a point where they'd be ready for the market, no one at the American Meat Science Association had seen any. That group is a national organization of scientists who contribute to the body of knowledge about meat, including how muscle grows and develops, and to the measuring of meat quality, food safety, and consumer marketing. The association is a potential partner, trusted by existing interests, that can help better understand cell-cultured meat.

Of course, giving that association access to cell-cultured meat samples isn't a necessity, but it would be helpful to have an organization comprised of leading scientists familiar with meat actually examine and learn more about how cell-cultured meat compares with what's generally understood about the conventional meat we're all used to eating. Indeed, at a public meeting on the subject held in the summer of 2018 at the US Food and Drug Administration, Rhonda Miller, a past president of that association, said as much.

"Meat scientists do not have enough information about cultured tissue to determine whether it should be called meat or
how it should be regulated," she said. "Samples of cultured tissue
have not been available for evaluation of the safety, composition,
nutritional bioavailability, functionality, and sensory properties
to understand how it compares to meat from conventional animal production."

From a layperson's perspective, it's actually pretty difficult to
talk about *what* cell-cultured meat is exactly. It's not hard to
understand the basic principles. Sure, you grow fat tissue and
muscle tissue, put the two together, and what you've got is real
meat. But it's hard to speak authoritatively when talking about
a process that's cumbersome to describe beyond those simple
terms. To be sure, this is true for many processed foods—after
all, who knows what goes into a Pop-Tart? But meat is a fundamental pillar of the diet. What happens when consumers can no
longer understand how their meat arrived on their plates?

This acknowledgment underscores a striking aspect of the
rise of cell-cultured meat that lays bare one of the nascent industry's toughest challenges. There's a gulf separating what ca-
sual consumers can know about it and what high-tech scientists
with multiple degrees understand about it.

Back in the United States, California-based Memphis Meats
CEO Uma Valeti explains that a lot of open questions about cell-
cultured meat will be answered as the products near the point at
which they're ready for market.

Valeti, already an accomplished cardiologist in his own right, incorporated Memphis Meats in August 2015 after spending a decade thinking about using stem cells to grow meats.

"I researched every single thing I could get my hands on," he told me one sunny afternoon at his company's headquarters in Berkeley. "I just got more and more convinced, the pieces are all here and we just need to put them together."

From the beginning, he said that gaining consumer trust has been at the top of his mind. He said Memphis Meats couldn't operate in stealth mode because it was essential that the public conversation around cell-cultured meat happen publicly. The start-up has hosted tastings and has invited journalists to try its meats. And Valeti aims to release his product under its own branding—as opposed to licensing its technology out to a larger existing meat company brand—in order to be closer to the consumer experience, and to allow consumers to be as close as possible to the development experience. It's a philosophy that runs throughout the company, he said.

"Everyone who's joined us came here for a once-in-a-lifetime opportunity of doing something very meaningful and I think to show the experience of what we're building to a consumer completes the cycle," he said. "If we do it really well then opportunities open to scale up."

And in regards to communicating most effectively with consumers, Valeti believes it's all about getting product onto their dinner plates. That's the best way to answer questions about whether the food tastes good, is safe, and is nutritious.

"Once people start tasting it, that's where the magic happens," he said. "I don't see much of a point in fighting and saying the food is safe. We definitely have to follow the guidelines of the agencies.

"This has got a production process that's a lot more traceable than the current industry," he continued. "It's got inputs that are traceable. The entire production is being done in one place. There are lots of safety checkpoints along the way and then you have a nutritional analysis that you can look at."

My discussion with Valeti is interrupted when one of Memphis Meats' food scientists brings out a plate on which, next to a few gooseberries and capers, sits a sample of the company's duck. It's been pan-fried, seasoned only with salt and pepper.

I look down at the product and marvel at how similar it is in appearance to a chicken tender. It's impossible for Valeti to mask his enthusiasm. He leans forward as I pick up the tender to examine it more closely.

Valeti urges me to pay special attention to its texture and appearance. "You can eat it by itself, but I would really encourage you to pull it apart, smell it, and check out the fibers, see how it will hang just like meat. Those are really important things to get."

I'd never tried something this advanced. Months before JUST had served me its duck pâté and duck chorizo, but this was a full muscle that I could hold in my hand and interact with by pulling it apart. I pick it up and take a bite.

Ten full silent seconds pass as I experience it, pulling it apart with my fingers.

"Oh wow," I exclaim, noting the fibrous nature of the meat.

Imagine pulling apart a chicken breast. When you do that, you can see that chicken meat is actually quite stringy. When you try to replicate that quality in a laboratory, though, it's really hard. The scientists at Memphis Meats have had to get the right liquid medium and bioreactor technology down in order to encourage the cells to mimic that process without the use of complex scaffolding. But they'd done it, and I was holding a prototype of it in my hands.

"This is a one hundred percent cell-cultured meat product?" I ask.

"Correct," Valeti says.

"This is wild," I say. Truly, it's the most convincing piece of cell-cultured meat I've seen or interacted with.

"Everything you would expect in meat is in there, and it's all cell-cultured," the scientist tells me.

"That is so cool," I say, still bewildered—even humbled—by the experience.

"You would never get this experience with non-animal-based products," Valeti says, meaning plant-based meat substitutes like Impossible Foods and Beyond Meat.

Valeti's confidence in his product is noticeable. "One of the things we do with all the tastings is we try not to put anything on the meat," Valeti explains. "I know for sure if you put this in the hands of one of our chefs and said, 'Okay make this into a very gourmet meal,' they could do it. But the key to this for a tasting is people should taste the actual substance. So that's the reason why we've only put salt and pepper on it."

I walk away from Memphis Meats headquarters in a daze, feeling as though I've brushed up against one of the rare, jagged, and unvarnished edges of the future. The work being done around the world is impressive, especially as it marches into fruition, materializing more and more into the types of products that I just picked up and tasted for myself.

All of the jostling entrepreneurs are united by a dream. All of them want to see that dream realized. They want to buy and taste cell-cultured meat. They want to see the world changed in the name of one of the most noble reasons possible—to conserve and be good stewards of the Earth. But in spite of those dreams not everyone in this young, global collective of companies is on the same page. Some products are scientifically more impressive than others. Because of this, the start-ups eye one another with suspicion, curious about who will fire their meat into the market first and how it will be perceived.

The big question that was emerging, though, was just where in the world this new meat would be introduced. Tetrick had boasted in June 2017 that he'd get his product to market first. He told me he'd have it in some restaurant or store by the end of 2018, though he hoped it'd happen months before that. But then, as he watched his own ambitions to release in the

Netherlands crumble, it became very clear to the people at JUST that government regulators—not scientific advances or technological prowess—would be the great decider of which society would taste cell-cultured meat first. And nowhere on Earth has that regulatory debate played out more fiercely than in the United States.

10

THE BELLY OF
THE BEAST

A short man with curly hair steps forward and interrupts my conversation with Tetrick, leaning down and speaking softly into his ear.

"I know you don't, but it's at ten a.m.," I hear him say.

"Are we seeing their slaughterhouse?" Tetrick asks.

The buzz of the JUST office is at its usual steady hum. I lean forward a little, straining to hear them better.

"They can take us to a slaughterhouse."

"I want to see one," Tetrick replies, staring at the man with an intensity that surprises me.

"We can ask them."

"He said I could before."

"I'll text them," the man says. "I mean, I don't want to disrupt it, but I'll send him a message."

"Okay," Tetrick says with a nod, swiveling around to meet my quizzical expression. Seeing my confusion, he fills in the blanks.

Hernan Jaramillo is charged with business development at JUST. But any title JUST might give him would be a buttoned-up way of saying the man has a very cool job. On any given day, Jaramillo is very likely on another continent, chatting with business executives—often over rounds of stiff drinks—about the opportunities to be had by partnering with his start-up. As JUST has pushed vigorously into the cell-cultured meat space, his job has largely revolved around connecting with meat companies about its work.

For weeks Jaramillo has been trying to broker a meeting between Tetrick and executives at JBS in Brazil. JBS is the world's largest processor of fresh beef and pork, and it does north of $50 billion each year in sales. If such a meeting was to happen, it would be a very, very big deal, and open the door to a major partnership.

Tetrick is looking for more than just a financial investment. He wants JBS to use his technology.

And this is one area where I think JUST stands out against some of the other start-ups in the cell-cultured meat field, save for Future Meat Technologies. By playing with a business-to-business model, a company can potentially scale the reach of its product dramatically, though there are often some concessions over control.

It's one thing for a company such as Tyson Foods or Cargill to invest money in a start-up; it's another thing entirely for them to redirect that investment back to themselves, agreeing to rejigger their own internal processes to make cell-cultured meat on their own. And that's the sweet spot that Tetrick is going for.

"You can have a bunch of little investments—two million or three million dollars for these guys is nothing," he says. "But what about investing in infrastructure?"

He paused to clear his throat.

"I want them to fucking buy bioreactors, you know?"

And this is the start of Tetrick's newest dream for his company. He's not out seeking investment dollars anymore. He's looking to forge deals with existing companies. Big companies. Companies that either have financial heft or a certain cachet. He wants executives at JBS and Cargill and Tyson to go back to their conference rooms and talk about building a thirty-thousand-square-foot facility outside Rio de Janeiro, stock it with big bioreactors, and write a check to JUST, who'll supply them with the knowledge and licensing needed to grow their own meat.

"That's what I want," Tetrick says. "That, to me, are signs of the end [of conventional meat]."

Whether those deals will ultimately be sealed, and whether that type of licensing agreement proves to be the best and most effective path forward for cell-cultured meat products, has yet to bear out.

This strategy flips the traditional sense of vegan activism on its head—underscoring how, in short order, very intense vegan activists have become very real agents of change by altering their approach. A committed vegan like Tetrick looking to schedule a friendly visit to one of the world's biggest meat company's slaughterhouses? One of the world's biggest meat companies collaborating with a vegan like Tetrick? It's strange to think about, and even odder to watch play out before you.

This change in role is perhaps no more evident than in the growing influence of a man named Bruce Friedrich. He isn't running a cell-cultured meat company, nor is he employed by one—but his Washington, DC–based nonprofit, the Good Food Institute, is a kind of central nervous system for the nascent industry. It advocates for and represents the interests of cell-cultured meat companies (and plant-based foods) to the government and business. He's become a force, a go-to resource for Silicon Valley's food technology CEOs such as Tetrick, and an unapologetic fighter for the industry as cell-cultured meats have slowly inched onto the radars of food regulators in Washington. And he's just one figure among several looking to support this new-age animal rights movement, one piece in a mosaic of support that includes longtime activists and deep-pocketed vegans. Bump into Friedrich and you'll likely find him in a neatly pressed button-down shirt and clean shaven—but this wasn't always the case.

T hey probably rolled up in a Volvo. He searched his memory but couldn't say for sure. It was chilly—three-o'clock-in-the-morning chilly. But yes, that was probably the type of car. It was also really dark; after all there was just no way they could do this before nightfall. But yes, Volvos are the chosen vehicle for

the radical brand of Catholic leftists who protest nuclear proliferation. It was probably that.

But it was definitely chilly.

When the car rumbled away, Friedrich, then twenty-four, was
stranded with three others, leaving them southeast of Raleigh,
North Carolina, in the woods just outside Goldsboro. It was December 7, 1993; Tetrick would have been thirteen years old at
the time. In the clear sky, a canopy of stars acted as a heavenly
compass for the quartet as it plunged north in a slow march toward the 3,300 acres that make up Seymour Johnson Air Force
Base, home to the F-15E Nuclear Strike Eagle. They stepped forward with hushed voices, murmuring in prayer as they pressed
into the early morning hours navigating three quarters of a mile
through a dense and remote woodland.

Unlike the used Volvo, the F-15E Nuclear Strike Eagle is a
$34 million two-seater, a high-speed killing machine designed
in the 1980s by McDonnell Douglas (now Boeing) specifically
for ground attacks. They were eventually outfitted with laser
designators to assist in the precision firing of guided bombs, including the kind that could wipe out anyone within a three-
kilometer radius of a one-megaton nuclear airburst—a bright
flash of radioactive light followed by deathly quiet.

Friedrich was accompanied by Philip Berrigan, a seventy-
year-old patriarch of civil disobedience; John Dear, a newly or-
dained thirty-four-year-old Jesuit priest; and Lynn Fredriksson,
thirty, a homeless-rights activist from Baltimore. They called
themselves "Pax Christi—Spirit of Life Plowshares."

The Washington Post would later report the incident as the first case of political vandalism at the air force base. A spokesperson for the military, Jay Barber, would call it a publicity stunt and a waste of taxpayer dollars.

"If they had been out to destroy a jet, they wouldn't have brought their hammers with them. They would have brought dynamite," Barber told a reporter.

The four activists had spent nearly ten months preparing, which according to Friedrich consisted of some research, but mostly spending time with one another to "become family." They shared their life stories, prayed together, scoped out the geography of the base, and even paid a visit to the Smithsonian's National Air and Space Museum where they bought a book about the fighter jet that included a detailed diagram of the plane.

On the night of the action, they had just a few hours to complete their mission. A press release about their peaceful activism and subsequent arrest had already been written and arranged for release at six o'clock in the morning, by which time they assumed they would have completed what they set out to do.

"Nobody had a cell phone, this was 1993, so there was no way for any of us to connect with anybody," Friedrich said.

They crawled up a dark knoll, cresting a hill and peering out onto a sight that stopped them cold.

"Fuck."

The entire base was flooded in light. Thousands of people were on the field between the activists and the jets. The base was conducting a war games exercise, meant to simulate aspects of warfare at the tactical, operational, and strategic levels. The four activists sat there at the top of the hill just staring. They looked like hippies, draped in modest plainclothes and long hair. This was never supposed to happen.

A moment passed by. It was loud down there. Another moment. The Volvo was gone, there was no going back. Another moment.

"Fuck it!" Berrigan said. "We're going."

They jumped the back fence of the base and proceeded into the chaotic scene. And much to their bewilderment, no one noticed. People were running, yelling, and following orders—seemingly unaware and undeterred by four obviously-out-of-place activists.

"Nobody said a word," Friedrich recalls. "We just walked up to the plane and started doing our little hammering thing. And then, at that point, that drew attention to us."

One strike at the plane's ultra-high-frequency antennae and the hammerhead went flying and the handle shattered. It looked like something from a cartoon, Friedrich recalled. With just a hammer they caused an estimated $25,000 to $30,000 in damage.

It took just seventy-five seconds, according to newspaper accounts. At first there were looks of shock and surprise. Then the four were detained. Around them, officers barked into their walkie-talkies, "This is real world! This is real world!"

All four were charged with willfully damaging federal property, setting each of them up for a potential sentence of ten years in federal prison and a possible $250,000 fine. They appeared before US District Court Judge Terrence William Boyle and admitted to what they were accused of.

Judge Boyle released John Dear and Philip Berrigan after nine months in prison. Friedrich and Fredriksson remained in prison for another year. Twenty years later, in a letter published by the *National Catholic Reporter*, John Dear wrote about his experience:

"I didn't enter the Society of Jesus to sit around and drink or watch TV with other Jesuits, but to follow Jesus, who was killed for doing civil disobedience in the temple," he wrote.

Friedrich brings that same conviction to everything he does. When he was younger, he initially focused on the issues of homelessness and global poverty. He became vegan in 1987, six years before the incident at the military base, after reading *Diet for a Small Planet* by Frances Moore Lappé. The book was the first major one to critique meat production as wasteful, environmentally unsustainable, and a contributor to global food scarcity. His reasoning for the switch actually had nothing to do with animal welfare, as much as it did with people going hungry around the world. As he explains it, a person gets one calorie for every nine calories of feed that goes to animals. For Friedrich, it seemed akin to taking food directly out of the mouths of people who need it most.

"Just the idea of cycling crops through animals so we can eat animals just struck me as patently immoral as part of my Christian faith," he says. "It's morally equivalent of throwing that food away."

Later in life he would read and be influenced by *Christianity and the Rights of Animals* by Andrew Linzey.

"And then there was a confluence of factors," he said. "Suddenly I saw what was happening to animals as the most important thing from a perspective of Matthew 25—that other animals are loved by God just as much as human beings are, and a vast majority of people don't think about them, including Christians."

He worked with PETA, which led to introductions to Josh Balk and other influential people who would later show an interest in cell-cultured meats, which helped Friedrich embrace a new path of resistance. His own personal evolution took place over the course of several years of intense activism, but the core of his experience is mirrored by so many others in the field. They start as relatively hardcore activists who are critical of the broader system, and then gravitate toward an ideology that uses capitalism to subvert a part of the system they hope to disrupt. And those people are of every stripe. Some are scientists. Some are lobbyists. Some are entrepreneurs. Some are even venture capitalists.

So as Tetrick nurtured his company from the ground up, tackling animal welfare issues from a decidedly corporate sphere,

Friedrich would go on to create the Good Food Institute, which would arguably become the central hub for the larger movement.

T o be sure, not every influential vegan activist expresses themselves with the same personal intensity as Friedrich. As more companies enter the space—there are more than thirty now—they are all seeking funding, and lucky for some of them, there's a coterie of vegan investors eager to step in as early-stage supporters.

The initial investment firms looking into cell-cultured meats include Stray Dog Capital, Clear Current Capital, and New Crop Capital, among others. Vegans are in the executive ranks of nearly all of them.

"Rather than try to veganize the world—which I don't think can happen—going from the top down just makes sense to me," venture capitalist Curt Albright told me. "It just seems like everything is prime."

Albright retired in the summer of 2017 after twenty-six years working as an investment banker, many of those years spent in Charlotte, North Carolina, at a firm called Vining Sparks. A numbers guy since childhood—he was tracking stocks in *The Wall Street Journal* as a kid—Albright was eager to pour his time and energies into the cell-cultured meat movement. So he approached Friedrich and asked how he might serve. Fast-forward several months and Albright founded Clear Current Capital, a venture capitalist firm looking to enlist investors who can contribute no less than $250,000 to put toward fifteen to twenty new companies. Among the cell-cultured meat ventures, Albright admits he's most excited about Memphis Meats and Finless Foods, a start-up focusing on seafood, particularly blue-fin tuna.

"I look at this as oiling the machine," he told me. "We need capital and we need liquidity. The more funding we get it's all going to help the machine grow."

And what would the vegan funding world be without its own special groups? These moneyed heavyweights started a group called the GlassWall Syndicate—a council that brings together venture capitalists, nonprofits, trusts, foundations, and individual investors. Lisa Feria, the president of Stray Dog Capital, created the syndicate.

Feria started her career at General Mills, literally making brownies and Cheerios, brands that sell in virtually every US supermarket. Then she went to the Procter & Gamble Company, where she worked in consumer goods. This was also about the time she became vegan. She operates her investing from the perspective that animals need to be removed from global supply chains. Whether that means growing leather from cells, funding start-ups that create non-animal-based gelatin, or pumping money into a cell-cultured meat start-up, Feria knows that interest from investors will be crucial if these start-ups ever want a shot at challenging the multibillion-dollar status quo.

"My level of contribution to these companies comes from three areas: capital when they need it to grow, access to our networks and contacts, and really just trying to figure out areas of consecutiveness between other companies," she says, noting that she typically invests when start-ups are in their earliest stages. "We prevent these companies from going under way before they show their potential."

New Crop Capital was founded in 2015 and is run by a vegan named Chris Kerr, whose footprint in the food technology world can perhaps best be summed up with the headline of a profile written about him in *Bloomberg Businessweek*: "THE VEGETARIANS AT THE GATE." It was in that piece that Kerr cogently expressed what has become obvious in the last several years, that the vegan revolution is upon us, "and there are fortunes to be made."

"We as a species have locked into a nutrient platform that, as it scales, was destined to fail. We're so locked into this form of putting nutrients in our body, it's really hard to break that," Kerr explained.

But to be produced at scale these new vegan foods will have to be manufactured with some of the same complex processes that currently feed the world, and to go global this new crop of start-ups will need money. And a lot of it. The fund manages $25 million and has invested in Memphis Meats, Beyond Meat, and New Wave Foods, among others. Friedrich is its managing trustee.

It's not just vegans looking to fuel this nascent industry, though. Seeing the potential, plant-based and cell-cultured meat companies have garnered the attention of some of the biggest names in both food and finance. Goldman Sachs Group in

January 2018 pitched money into what ultimately wound up being a $65 million investment round for Ripple Foods, which makes a milk alternative with yellow peas. Impossible Foods, the maker of the plant-based Impossible Burger, has funding from UBS, Bill Gates, Singapore's sovereign wealth fund, and one of JUST's initial investors, Khosla Ventures. Tyson Foods, Cargill, and Richard Branson have invested in Memphis Meats, in which Kerr's group had already invested.

It's an uncommon situation. Big Money and vegan activism have collided and have found a common rhythm. A purpose to drive mutual benefit. Industrialization, in many ways, has wreaked havoc on laborers, made humans less healthy, and arguably made the world a dirtier place. Now, with these two interests joining forces, an optimistic viewer could deduce that there's a big opportunity to make things better through that shared vision that was originally forged in the image of veganism.

But there is still a palpable sense of anxiety buzzing beneath the surface of the enthusiasm. Investors and activists alike often talk about the development of cell-cultured meat and other meat alternatives as if they are in a race against time.

They are. We all are.

The United Nations in November 2019 all but admitted that the goal to limit global temperature rise to 1.5 degrees Celsius will practically be impossible. That's not good news, as scientists have determined the effects of any warming higher than that would be disastrous for life on Earth.

"There is no sign of [greenhouse gas] emissions peaking in the next few years; every year of postponed peaking means that deeper and faster cuts will be required," the UN report said.

Just how big a difference would a rise of 1.5 degrees Celsius be compared with, say, 2 degrees Celsius for us? Huge. If humans delay serious action and allow warming to rise to that level, there would be an expected average of four inches in global sea level rise. Mass extinction of many plants and invertebrate species would take place, and coral reefs would all but disappear.

"The livestock industry is [horrible for] greenhouse gases and water," said Sonny Vu, the entrepreneur and investor based out of Singapore who introduced Tetrick to his future Hong Kong financiers. "This is unsustainable. So there must be sustainable protein sources. We've got to figure out how to make twice as many cows and three times as many pigs with the same amount of land or less. It's not going to be by shoving more into mechanized farms."

Vu openly admits that the wearable technology business—which he was in when he first started getting nerdy about food tech—is not his primary passion. This is why he's made a habit of routinely diving into extracurricular passions, which includes marveling not only at the scientific advances made possible through food technology, but also at the potential impact this crop of cultured meat start-ups might have on the climate.

"I'd never encountered companies like this," Vu told me.

He's got a survivalist streak. Vu recalled a conversation he had while visiting with a climate research scientist friend in the Bay

Area. As he sat down with Jack Sculley, then of the National Center for Earth Surface Dynamics at the University of California at Berkeley, he implored the researcher to give him a dispassionate rundown of several probable models of what the Earth would experience in coming years as a result of the climate crisis. The bleak response he got left him feeling stunned.

"I was like, 'Shit man, how can you say that and be as calm as you are?'" he recounted. "Maybe he was resigned to the fact that this was going to happen. I refuse to believe that. No. We are not past the point of no return."

He paused, considering the realities of the climate changes expected to come in the future. Thinking about it works him back into a state of worry.

"We are royally fucked," he said. "We need to change now or there will be no college for the grandkids. I don't know if I'm an alarmist or what not, but climate change is really, really real—and it's way scarier than people imagine."

In Vu's view, food technology is one of the clearest paths toward forging good foods while preserving the planet for future generations.

Vu and his high-dollar investor ilk weren't the first to recognize the potential of cell-cultured meats and to put some capital toward the idea. This is something that Ingrid Newkirk, the founder of PETA, was quick to tell me. She considered it a worthy endeavor years before, after reading about Willem van Eelen in a newspaper. But the idea didn't catch on among her vegan peers at the time. Still, PETA has in the past devoted some of its resources toward supporting early work to develop cell-cultured meat. Newkirk gives a nod to Tetrick, Balk, Friedrich, and other prominent vegans working in the space, but not without a semidismissive—if not downright playful—potshot. After all, it was under the umbrella of her organization that a lot of these people got their starts.

"The boys can go off, the dot-com people can take it, and we wish them well."

In this case, the boys have gone off to scour the globe for a place to release cell-cultured meat.

And just when was that going to happen? I asked Tetrick.

He smiled sheepishly. It wasn't a terrible question for him. After all, for a full year he said he'd sell cultured meat by the end of 2018. Sure, there were setbacks, but with the end of 2019 approaching fast, it was worth asking again.

"I think we can do it by the end of the year," he said, indicating that Singapore was the frontrunner.

I nodded, thinking to myself, *We'll see. We'll see.*

11

FOOD FIGHT

I t isn't hard to become overwhelmed by Washington's elephan-
tine bureaucracy—the complexities of policy are the province
of wonks and wordsmiths. Learning how to navigate that world
is a full-time job for policy journalists, lawyers, and lobbyists,
who are tasked with untangling the politics of competing interests, the high emotions, and the steady pressure that Silicon Valley has increasingly pushed onto Washington. Lawmakers are
increasingly challenged to reimagine a world that breaks from
the conventions of the past, molded in the likeness of new technological possibilities. That has been the case with artificial intelligence, privacy concerns, the labor and business implications
of the gig economy, and certainly the proposition of lab-grown,
cell-cultured meat.

Beneath the surface, the policy challenges around cell-cultured meat are many. But on its face, the two biggest questions seem simple: "Is this stuff real? And is it safe?"

These are two issues that, for a while, were grappled with mostly in private, as nervous ranchers and farmers watched with cocked eyebrows as more headlines and stories detailed the efforts of the companies pushing to make cultured meat a reality. In conversations with me, representatives of several of the major meat industry groups invoked the term "fake meat" to disparage the new technology. To them, it isn't a godsend in the global fight against climate change, it's a threat that could upend their way of life. And in many cases, the mere existence of cell-cultured meat as a concept is a threat to family businesses that have been passed down for generations.

In a signal that they'd heard enough, the US Cattlemen's Association in February 2018 filed a twelve-page petition to the US Department of Agriculture's Food Safety and Inspection Service (FSIS) asking the department to narrow the definition of "meat" to the flesh of animals that have been harvested in a traditional way.

As the fight escalated in 2018 and 2019, it struck me as odd that our national food policies would ever be based on an idea of what's considered "traditional." A standard of traditionalism implies that our food system somehow looks or operates as it did as recently as fifty years ago. It doesn't. Egg and meat companies may print pastoral farm landscapes on the front of their food packages, but the reality is that very few eggs or cuts of meat come from animals that experience any such life. What's traditional about gargantuan farm operations that keep egg-laying hens crammed into small spaces, that grind to death millions of male baby chicks because they don't lay eggs, or about the factory farms that are built to house more than six thousand pigs? And perhaps more important, what are the benefits of keeping a so-called traditional system that operates so brutally, and at massive scale?

"Methods of cattle production today would have been unthinkable to our great grandparents," wrote the Good Food Institute in its rebuttal to the ranchers' petition. "If USDA were to limit meat and beef terms to the flesh of cattle born, raised, and killed in 'the traditional manner,' almost no meat on the market today could bear such labels."

By the same token, cell-cultured meat companies themselves can come across as disingenuous. One of their primary marketing tactics is to try to convey that their products are the same thing as meat on a molecular level, that there is a comforting and allaying sameness to their meatballs, chorizos, and chicken nuggets that should somehow evoke feelings of familiarity. Even their names strive to communicate as much to the public. The bacon being produced by Mission Barns hardly involves a barn, Aleph Farms isn't growing meat on a farm, and more than two thousand miles separate the labs at Memphis Meats' headquarters a few minutes from downtown Berkeley from its namesake city in southwestern Tennessee, with its rich barbecue tradition.

In filing its petition, the US Cattlemen's Association decided it was time to demarcate the food products made by the current-day meat system and the ones being dreamed up in high-tech laboratories. The ranchers asked the federal government to box cultured meat companies out of the meat market by adopting a definition—which in Washington is called a "standard of identity"—that would specifically exclude them.

Historically, this isn't an uncommon tactic. The problem for the meat industry, though, is that it rarely works. The American egg industry was thwarted in its attempt to force Tetrick's JUST to stop selling Just Mayo as mayonnaise on grounds that real mayonnaise must include eggs. In the last decade, the dairy industry has waged several unsuccessful legal challenges against plant-based milks to try to force them to quit using the term "milk" on their packaging.

"You haven't 'got milk' if it comes from a seed, nut, or bean," Jim Mulhern, the president of the National Milk Producers Federation, said in response to a congressional effort to crack down on alternatives to cow's milk. "In the many years since we first raised concerns about the misbranding of these products, we've seen an explosion of imitators attaching the word 'milk' to everything from hemp to peas to algae."

Perhaps recognizing those industry losses, a second beef industry group, the National Cattlemen's Beef Association, submitted an official response to the petition a month and a half after the original was filed. Oddly, it asked the government not to adopt its sister group's definition. It was a striking moment.

The United States meat industry had almost always maintained a unified front over issues in Washington.

But in this case, the beef industry was essentially split. The US Cattlemen's Association said regulation should fall to the Food and Drug Administration (FDA), which mostly oversees dairy, seafood, produce, and packaged foods including plant-based imitation meat products. The other group, the National Cattlemen's Beef Association, said the USDA would be a better fit because it was already charged with regulating meat products as they go through slaughterhouses. And still others made the argument that it should be a combination of the two.

"If producers of lab-grown or cultured meat products wish to call these products meat, they must adhere to the same stringent food safety inspection standards and comply with the same set of labeling mandates as all other traditional meat food products," the National Cattlemen's Beef Association wrote to the government.

The cell-cultured meat companies, for the most part, initially said they would be fine being regulated by the FDA. The FDA has more working knowledge of the cell-culturing process by virtue of the fact that it already oversees the medical research and pharmaceutical industries.

Throughout American history, the introduction of new foods has been replete with this sort of drama. The tactics are now easy to spot. It almost always starts with an established group trying to discredit a new product by disparaging it publicly (ranchers calling cell-cultured meat "fake meat").

Then they attempt to use political clout to undermine the new product's ability to flourish in the economic arena. Discredit and disparage. Rinse and repeat. It's a playbook that's more than a century old.

In 1875, margarine makers first made their way across the Atlantic Ocean from continental Europe to try their luck in the American market. The spread had only been invented six years prior by French chemist Hippolyte Mège-Mouriès using beef fat as a primary ingredient.

Mège-Mouriès created the product in response to a contest announced at the 1866 Paris World Exhibition. Troubled by Otto von Bismarck's militarization of Prussia and by butter prices set far too high for a new wave of city dwellers, the French government had sponsored a competition for the best butter substitute, a cheap product that would be suitable for use by the lower class and in the armed forces. Mège-Mouriès would later sell his margarine patent to the Dutch company Jurgens, which is now owned by multinational food giant Unilever.

The substance was the subject of immense skepticism the moment it hit the US market, according to the Wisconsin Historical Society. In dairy circles, it was described as a fraudulent product, a "gigantic swindle of the age" that would be passed off to consumers as "the real thing."

Today we might consider butter and margarine innocuous products on the dining room table, but the turf war between the two groups was particularly significant in American history and would grow into one of the nation's most vicious political food fights. At the center of that fight was William Dempster Hoard, a dairyman who would go on to serve a single term as governor of Wisconsin.

Hoard's paranoia is noteworthy because it gave rise to the nation's dairy lobby. Fearing that the dairy industry might fade due to the threat of margarine, he founded the Wisconsin Dairymen's Association in 1872, one of the first of several state dairy lobbying groups. These groups set to work, getting members in front of statehouse politicians across the country. In 1881, Missouri implemented an outright ban on margarine. New York followed suit three years later, as did Maine, Minnesota, Wisconsin, Ohio, Pennsylvania, and Michigan. By 1886, twenty-four states had passed laws placing restrictions on margarine—five of them even decreed the substance would have to be dyed pink, a policy created because butter makers were furious that margarine producers would often dye their product yellow—never mind that butter makers often dyed their own product yellow, too.

That same year, dairy groups from twenty-six states convened in New York to settle on a case they would take to Washington, DC, to persuade congressional lawmakers not just to bring the hammer down on margarine, but also to formally define butter. They were successful. The Oleomargarine Act of 1886 passed; on August 2 President Grover Cleveland signed a bill into law that defined butter and levied a tax of two cents per pound on domestic margarine, fifteen cents per pound on imported margarine, and unleashed expensive licensing fees for margarine manufacturers, wholesalers, and retailers. Three years later, as governor of Wisconsin, Hoard would make his state the most aggressive enforcer of new anti-margarine laws by creating a new state agency, the Dairy and Food Commission, which was charged with inspecting and testing butter and cheese authenticity. To this day, Wisconsin law prohibits margarine from being set out on restaurant tables unless a customer specifically asks for it.

The conventional meat producers today are waging a similar battle over terminology and definitions. But instead of trying to force cell-cultured meat to look a certain way, their state lobbying organizations have gone from statehouse to statehouse—usually ones under the control of Republicans—seeking to persuade state lawmakers to pass restrictive rules around what the makers of cell-cultured meat can call their products—also the subject of fierce debate within the small cultured meat community. It's reminiscent of Unilever's tussle with JUST's vegan mayonnaise. Legalistic fights over terminology are often the first prong of the food giants' defense.

When Willem van Eelen first considered the technology, he called it "in vitro meat," a term that isn't particularly appetizing for folks pushing grocery carts up and down grocery store aisles. Tetrick says he'd like to just call it "meat." Others have toyed with "cell-based meat." In 2016, Friedrich made it a central part of his mission at the Good Food Institute to find a name the nascent industry could rally around.

He convened focus groups asking them their opinions of terms including in vitro meat, cultured meat, clean meat, lab-grown meat, lab-made meat, and others. The results did not paint any one clear picture as to what people preferred. Every term garnered its own skeptics. Still, Friedrich and others in the movement settled on "clean meat," which was meant to evoke the same sorts of ideas as "clean energy." If cell-cultured meat production pumped less greenhouse gas into the sky, and used less water and land, then it wouldn't be a leap to adopt the Earth-friendly nomenclature. It also helped communicate another positive attribute—though still unproven in practice—that growing meat in controlled, sterile bioreactors would curb the chance of bacteria festering. That meant a reduced threat of salmonella, *E. coli*, and norovirus.

The movement considered the term at the world's first cell-cultured meat conference in Haifa, Israel, in the spring of 2017. And while it wasn't met with overwhelming enthusiasm, the general consensus was that it was the best term that existed at the time.

So Friedrich and the Good Food Institute pushed it hard in meetings with corporate figures and with the press. Signs showed those efforts paid off. David MacLennan, the CEO of Cargill, the number two beef player in America, used the term "clean meat" during a live September 2017 interview with Maria Bartiromo on Fox Business. And Tyson Foods' then-CEO Tom Hayes referred to it in October 2017 on a food industry panel hosted by *The Wall Street Journal*. The term cropped up in articles produced by major publications, including the *Journal* and *Bloomberg*, legitimizing it in popular forums.

Even so, the cell-cultured industry still hadn't totally embraced it, and for good reason. Mark Post—a pioneer in the field—said he didn't care for the term because it didn't translate well into other languages, including his own native Dutch. It implied the meat had been run through detergent, he said.

Ira van Eelen also had doubts. The term "clean meat" is inherently judgmental. By calling a new product clean, it implied that the old product must be dirty. And that might turn people off.

"If you want to tell somebody to do something better, don't tell them that something that they are doing right now is dirty," she told me. "It's not fair because that's the only meat they have. Culturally, it's quite normal that people are eating meat. Be kind."

The coalition behind the term "clean meat" seemed to waver in the following two years, as meat lobbying groups in Washington sought to persuade the Trump administration to adopt policies that would make it more difficult for the start-ups to get their products to the marketplace. By this point, the number of cattlemen derisively using the term "fake meat" had increased, and it became clear the insinuation that their foods were somehow "dirty" had left them feeling put off.

Just as the dairy farmers projected their problems onto the makers of margarine, so too have American ranchers with cell-cultured meat. The ranchers don't concern themselves so much with the actual molecular makeup of cell-cultured meat—they instead showcase a stodgy fearfulness of its social and cultural connotations. That fear manifests in nearly every public jab they make toward the product, casting it negatively not for what it is, but for what it isn't. It's noteworthy that the beef industry did not ask the government to consider cell-cultured meat's nutritional profile or what it is on a molecular level. Those would have been substantive questions.

Maybe they didn't ask because it's not about the science. Maybe it's something more sociological. As put in a 1982 journal article delightfully titled "The Menace of Margarine," butter was an important symbol not because it was good to eat, but because it was "good to think." To the industry, butter embodied lush scenes of rolling pastures speckled with dairy cows—what was perceived by dairymen as "the good old days." This sort of thinking couldn't have been any more evident than in the words of Maine dairy farmer Harris Lewis in 1875.

"It is deplorable that the only perfect food for mankind, milk, or its products, should be adulterated to satisfy the unlawful and inordinate greed of a few soap-grease men, and thereby endanger the entire dairy interests of this country."

In that same vein, many modern-day ranchers have adopted that very attitude for themselves and their product. And so the playbook repeats itself, with an established industry pitting its ritually significant foods against a high-tech newcomer; a clash between a morally sanctimonious stalwart and a culinary interloper that's been stigmatized as a threat to the status quo.

The conventional meat interests have tied up the upstarts' time and money in fights over terminology. But even if they get everything they want—all the laws and restrictions and headaches for cell-cultured meat—will this strategy actually work to turn people off to the allure of these future products? History suggests probably not.

The dairy industry's reign over margarine lasted a good twenty years before its efforts began to unravel—at the hands of consumers, not policymakers. In 1924, people in Washington and Oregon voted by referenda to repeal prohibitions on butter substitutes. In 1926, voters in California repealed a two-cent-per-pound state tax on margarine. And in 1950, after a bitter fight by congressional lawmakers, the Oleomargarine Act was repealed and replaced with legislation that removed any tax or fee on margarine and its makers. In the late 1950s, per capita consumption of margarine officially passed that of butter, and its popularity continued to grow until the late 1970s. It wasn't until the mid-2000s that per capita margarine consumption would again dip below that of butter.

There's one clear reason established industry groups have failed to beat back new technology each and every time—with margarine, plant-based milks, and increasingly plant-based meats. Consumers are interested in trying new things. Sometimes it's because they have a financial reason (margarine was cheaper during periods of economic depression), sometimes it's for health reasons (plant-based milks have different nutritional profiles from conventional milk), and sometimes it might be for ethical reasons (cell-cultured meat is better for the environment and kinder to animals).

Cell-cultured meat isn't without its hang-ups. It challenges many of our basic sensibilities about how we should frame our diets, forcing unavoidable questions about how closely we should try to stick to the natural world and whether it's wise to place our fates in an ultraprocessed foodie future. I've met many people who have a viscerally negative reaction to the concept—but many, many more who are eager to try it for themselves.

Even as butter makers waged their war on margarine, historic data show consumption of margarine continued to skyrocket. In the six years after the US Congress adopted the Oleomargarine Act, the number of margarine manufacturing plants might have decreased from thirty-four facilities to about seventeen, but in that same period, consumption increased and continued to increase for the next several decades. In 1886, per capita consumption was about half a pound per person each year. By 1920, it was at about three and a half pounds per person each year. By 1970, that number had leaped to more than ten pounds. All that is to say, if consumers are hungry for a certain product, time has proven that no single industry—even with the help of the federal and state governments—can stifle that enthusiasm.

"I do know that the [cattlemen's lobby] is nervous about alternatives," said Ken Klippen, the former president of the National Association of Egg Farmers. "But I think people are afraid of change. Change is coming. I've been in this industry for forty years, and in the last five years I've seen more change than in the thirty-five previous years I'd been in it. We have to learn to bend."

The cell-cultured meat players have also gotten creative and found heavyweight allies. The ranchers might have resorted to an old playbook, but the rest of the meat industry, including the massive processors, wasn't totally on board. Memphis Meats announced an investment partnership with Cargill in 2017 and with Tyson Foods in early 2018. Around that same time, Israeli start-up SuperMeat announced it was getting investment dollars from PHW Group, the third-largest poultry producer in Europe. Mosa Meat courted and inked a $2.2 million investment deal with Basel-based Bell Food Group in July 2018. And Tetrick has continued his flirtations with JBS. It was a sign that the major players were at least interested in what the high-tech start-ups had to offer—and it was a clear departure from the failed tactics of yesteryear.

Then the rhetoric from those major meat companies began to change. No longer did they consider themselves meat companies, but instead "protein companies." It was a clever move, and perhaps one that will keep them ahead of a consumer-driven movement. By the time that language had been adopted, Los Angeles–based Beyond Meat had successfully gotten its plant-based burgers into mainstream American grocery chains, where its product sat displayed in literal meat cases, next to conventional steaks and ground beef. The success of Beyond Meat, and competitor Impossible Foods, marked a sea change for the American meat industry. One of the first inklings that major meat companies might actually embrace alternative proteins occurred in March 2017, when Tyson Foods CEO Tom Hayes appeared on Fox Business and publicly acknowledged that consumer demand was pushing his company to change its thinking.

"Protein consumption is growing around the world—and it continues to grow," Hayes said. "It's not just hot in the US; it's hot everywhere, people want protein, so whether it's animal-based protein or plant-based protein, they have an appetite for it. Plant-based protein is growing almost, at this point, a little faster than animal-based, so I think the migration may continue in that direction."

At the time, those words coming from the lips of a top meat industry executive was very much the exception and not the rule. And Tyson Foods was putting its money where its mouth was. By that time, it had acquired a 5 percent stake in Beyond Meat. These sorts of changes within the larger meat industry were underscored by the fact that grocery stores, where consumers interact with meat most directly, were at the beginning of their own meat alternative revolution. The meat sections of stores were quickly being recast as protein sections, opening the door for all sorts of new, nonconventional meat products to scramble out of the specialty foods section and into meat cases.

One thing that I think is important to point out is that the dairy industry ultimately did win its fight against margarine, though not through aggressive countermarketing or by lobbying the government to interfere in the market. No, margarine fell out of favor with consumers because tubs of the stuff were laden with partially hydrogenated oils to make it more shelf stable. In the 1990s, partially hydrogenated oils were linked to health risks including heart disease, stroke, and diabetes. This explains why the margarine my parents used to keep in the refrigerator during my youth suddenly vanished, replaced by butter. Data from the Department of Agriculture show that margarine consumption in America peaked in the late 1970s at about twelve pounds per capita, then plummeted to less than four pounds per capita by 2011.

Knowing what we know now about cell-cultured meats, I don't think this new technology will follow the same fate as margarine. Still, it's worth noting that in their first punches against cell-cultured meat, the beef industry didn't question the long-term health risks of the high-tech product. That seems important because, while cell-cultured meat manufacturers market their product as "cleaner" and as having an identical nutritional profile to the meat it mimics, for all of 2018 and 2019 it was impossible for the beef industry or consumers to know if that claim was true. We take cell-cultured meat start-ups at their word. To be sure, the long-term effects of the product—if there are any— just aren't known yet. It may very well go the way of margarine.

The debate over regulation highlights just how reactive governments are to technological innovation. JUST had at least two products—duck chorizo and foie gras—ready for market at the end of 2017, and Memphis Meats was already hosting small-scale tastings of beef meatballs, but because the USDA and the FDA had not yet ironed out a regulatory pathway for those products to enter the marketplace, they were delayed. To work through those delays, Silicon Valley's start-ups began hiring outside lobbyists to help them understand the capital's byzantine policy apparatus. JUST turned to Stuart Pape, who works for a DC lobbying firm called Polsinelli, in which he represents clients who need to navigate regulations typically imposed by the FDA and USDA. Prior to Polsinelli, Pape had served in a few positions in the Office of the Chief of Counsel at the FDA, including as associate chief counsel for food. He also had experience working as an executive assistant to former FDA commissioner Donald Kennedy. According to Pape, there really wasn't a clear, easy way to engage the government on the subject of cell-cultured meat in the first place. It was always going to be an educational battle.

"There's no established process for getting them to move along and work with you to create some regulatory pathway that doesn't already exist," Pape told me. "It's not like you have a meeting, and then you give someone a document, and then they comment on the document."

The process is like a mating ritual—Tetrick starts by singing the technology's praises, he and the government agency dance around the subject, the government makes a flirtatious but hesitant retreat, more dancing ensues, and finally some incremental movement happens. Preparation usually just sputters along at a snail's pace. Progress winds up happening out of necessity, rather than careful premeditation. That moment can get messy, like "six cabs arriving at the same Manhattan intersection at the same time," Pape said.

Ultimately, the move by the National Cattlemen's Beef Association to champion a position that would have the USDA regulate cell-cultured meat was both inspired and insidious. In advocating that cell-cultured start-ups be regulated by the same government agency overseeing conventional meat, the cattle-men essentially offered Silicon Valley's technologists what they wanted all along: equal footing for their high-tech meat. But it might also have been a regulatory trap. The USDA has long been criticized for its dual role of regulating and promoting agricul-tural industries. Some cell-cultured meat players expressed to me the fear that they would not have the same degree of influ-ence within the USDA as conventional groups, and so, under USDA regulation, they would be forced onto a lopsided playing field. It would also give conventional meat companies a seat at the table in figuring out how oversight of cell-cultured meat should go about—something nobody in Silicon Valley particu-larly wanted.

"I don't want the US Cattlemen's Association or the NCBA to back us into a corner," the Good Food Institute's Jessica Almy told me. "I think they are trying to take control of this narrative and it's not their story."

Pape thought the USDA would have a difficult time making the case that it should be the primary regulator of cell-cultured meat. The bulk of the responsibility should fall to the FDA, he argued, pointing out that the USDA's historic expertise doesn't relate to cell-culture production processes.

"This isn't observing the health of animals when they are brought into slaughterhouses, or how they're slaughtered. You can make lots of comments about whether FDA can keep track of all the foods, but this is just domestic protectionism."

Pape set to work trying to figure out an alternative. In his estimation, the way forward for cell-cultured meat companies wouldn't be to slide into any existing regulatory framework. The best way forward would instead be to work with the government to create a new process specifically designed for cellular agriculture food products.

It would be a difficult journey, Pape said, but worth it. Regardless of how that effort unfolded Pape said he knew one thing for certain: Going to market without the FDA's approval would be a misstep. Technically, the FDA does not have to approve food, drinks, or dietary supplements before they go to market, though companies are required to register their products with the FDA. Registration is not indicative of formal approval. Taking the shortest, easiest route to market is seen by nearly every company in the cell-cultured meat space as a bad idea because— right or wrong—it would communicate to everyday eaters that the companies don't care about food safety, even if the companies themselves felt confident about their product. Any seed of consumer mistrust could wind up creating a public relations nightmare for the whole industry.

"That would be the dumbest strategy that you could possibly follow," Pape said. "I don't think these are commercially viable if you haven't had some form of regulatory interaction."

By creating a distinctive new regulatory path for cell-cultured products, Pape hoped to strike a careful balance, one that would ensure the government gets all the data it needs while also allowing the companies to remain nimble enough to attract investment and allow for innovation.

"If you go too far in either direction you screw the pooch," he said.

In the meantime, Pape was keenly aware that meat groups could play the ultimate wild card.

"We haven't touched on what Congress might do," he told me. "The agricultural interests have enough stroke in the Congress to get some outcome here."

Getting traction with policymakers inside the FDA and USDA is already a really difficult task, and it occurs beneath the umbrella of the executive branch of government. What makes a lobbyist's job so tough is that they also have to keep an eye on the whims of congressional lawmakers, who can be persuaded by interest groups to take action. That could include something as big as creating legislation to dictate how cell-cultured meat would be regulated (a very unlikely scenario) or tacking obscure amendments on to large pieces of legislation that would shape how the FDA and USDA approach regulation.

Seriously courting lawmakers would mean cashing in political chits, but if a major meat lobbying group felt its concerns weren't being reasonably heard or addressed inside the agencies, appealing to Congress would be an option.

At the time, there were a lot of moving parts to consider, and the situation felt particularly precarious for cell-cultured meat companies going into early 2018. American agriculture had been taking a beating from President Donald Trump's escalation of trade wars around the globe. These skirmishes had inflicted lasting damage to agricultural profits, putting much of the industry on edge. But the meat groups held sway with the president and his Republican Party. Under enough pressure Trump might throw them a proverbial bone, directing the USDA to take full regulatory authority over cell-cultured meat.

But then something surprising happened. The FDA stepped in the way.

On June 18, 2018, the agency's then–top official, Scott Gottlieb, issued an announcement saying that increased interest by consumers and companies in cell-cultured meats merited a public hearing on the topic—and that his agency was going to host it. There was no mention of the USDA; no indication that the USDA was ever taken into consideration at all. It ruffled a lot of feathers.

The move was in line with language that the FDA's Susan Mayne had delivered a couple months prior at a conference in San Francisco at the Future Food-Tech Summit. At the time, she was the director of the FDA's Center for Food Safety and Applied Nutrition.

"I encourage you to work with FDA," she told entrepreneurs, "to bring your brainpower and imagination to the challenge of feeding the world, helping people have healthier diets, and providing innovative new options for consumers."

That the FDA was making a play to assume regulatory authority over cell-cultured meat wasn't entirely a shock. After all, the agency already regulates some 80 percent of the food supply, including most fish, produce, and some specialty meats. The USDA, by comparison, regulates mostly meat products and catfish. To further boost its position, the FDA asserted that its power over cell-cultured meat could be drawn from the Federal Food, Drug, and Cosmetic Act.

The maneuvering added a surprising new contour to the larger debate. It undercut strategic moves that were being made by meat lobbying groups and transformed a wonky policy discussion into an all-out interagency turf war.

"This is definitely surprising," Danielle Beck, the director of government affairs for the National Cattlemen's Beef Association, told me at the time.

Watching from the sidelines in Silicon Valley was nerve-racking for cell-cultured meat companies. Still, they were heartened by the fact that the FDA was stepping up for them. It offered a glimmer of hope for the effort to create the regulatory regimen they wanted all along.

"The leaders of the national and global meat industry want to feed the world animal protein in a sustainable way," Tetrick said. "That's a shared interest that should be celebrated. And I'm grateful to see the FDA bringing all of us together to talk about making it happen."

There was one big problem, though. Officials at the USDA were sour about the FDA's announcement. Not only were they excluded from the FDA's public hearing, but the FDA wouldn't acknowledge that they had any role to play in the regulation of cell-cultured meat at all.

The friction between the two agencies was worrisome. The more extreme the internecine fight between the FDA and USDA got, the harder it would be for the cell-cultured meat start-ups to champion the smooth creation of a new regulatory process for their products.

Seeing two of its agencies squabbling publicly over which would regulate this new technology, the White House intervened, summoning leaders of both to a joint meeting behind closed doors on June 22. Little was resolved in that meeting, though, as shortly afterward, the House Agriculture Committee, very likely at the request of a meat lobbying group, sent the White House a letter of complaint.

"Our expectation was that the White House meeting would direct USDA and FDA to coordinate on this issue," it read. "Moving forward, we would request that as the administration continues to evaluate these technologies, USDA and FDA be required to work together."

Two weeks later, at the New Harvest conference in Boston, the cell-cultured meat industry eagerly awaited a presentation by Eric Schulze, a senior scientist at Berkeley-based Memphis Meats. That the start-up hired Schulze was a prescient move by Uma Valeti, as Schulze had spent six years working as a federal regulator within the FDA with a focus on novel biotechnology. It's as though he was engineered for this exact time and place. He had an insider experience with government, so his insights were highly anticipated. His message to his peers at the conference was simple. They needed to buckle up for the road ahead.

"This is just the beginning of what is sure to be—and should be—a very thorough conversation about our products," Schulze had said. "It will take an investment on our part. An investment in our time and patience and continuing to place a high priority on transparency."

In the following week, the FDA made clear it wasn't backing down from the fight.

"This is not our first rodeo, so to speak, in this area," Mayne had said, adding that the agency was already planning a fall meeting on cell-cultured meats with its science board.

Clearly riled, the meat industry coalesced and on July 16, 2018, sent a letter to the White House, appealing directly to the president to intervene. It was one page in length and signed by the so-called DC Barnyard, which includes seven powerful trade groups representing the interests of an agricultural industry that contributes roughly $1 trillion to the American economy each year.

"If cell-cultured protein companies want the privilege of marketing their products as meat and poultry products to the American public, in order to ensure a fair and competitive marketplace, they should be happy to follow the same rules as everyone else," the letter stated.

The next month, major news broke in the cell-cultured meat space. Schulze, at Memphis Meats, was able to negotiate with one of the more powerful meat lobbying groups, North American Meat Institute, which represents the big processors, including Tyson Foods. On August 23, the two entities sent the White House a letter asking that the president move forward with a proposal they'd drawn up, which they believed could settle the controversial debate over how cell-cultured meat products should be regulated. The proposal would give the FDA and the USDA joint oversight.

Under the plan, the FDA would be responsible for ensuring that cell-cultured meats undergo premarket safety tests. The USDA would be responsible for potential continuous monitoring of cell-cultured meat-processing facilities, in much the same manner that it already oversees the processing of foods such as chicken nuggets and hot dogs. The USDA has inspectors in more than six thousand slaughter and processing facilities in the United States.

The letter did one other thing that proved controversial. It introduced the seemingly meaningless term "cell-based meat" to the regulatory vernacular, putting to bed the term "clean meat" and resolving one of the more contentious aspects of the product. The idea was that a truce would de-escalate the fight between the two agencies and potentially be the start of a good-faith effort by both Silicon Valley and established farm groups to move forward with a truce.

It worked.

On March 7, 2019, the two government agencies announced they'd signed a formal agreement to jointly oversee the production of cell-cultured meat. That began the long process of ironing out the wrinkles and working with all parties involved to figure out the finer details of what regulations would look like in practice.

For most of the cell-cultured meat community, the announcement felt like an important step forward. Certainly one thing was clear: In the United States, Schulze and Memphis Meats were leading the way, proving that they had the relationships and chops to sit at the table with established meat interests and hammer out a deal.

And even though Pape's vision of creating a new regulatory pathway for cultured meats had been stymied, Tetrick was receptive to the truce that had been drawn up. Throughout the fighting in Washington, he had seemed detached from the drama, preferring to let others maneuver in his stead. He'd been spending more time traveling outside the United States than inside it. His eyes were trained on another prize altogether.

12

PROMISE ABROAD

All the noise and excitement left an early impression. Every eyeball in the room was trained center stage, right on Tetrick, at the glitzy Tencent WE Summit in November 2016 in Beijing. Dressed in basic blue jeans and a gray sweater, he paced back and forth before turning toward a crowd of more than a thousand people.

"This is the question that has obsessed me for the last five years," he said. "What does it look like if we just start over?"

It's a question he's asked investors and audiences over and over again when he talked about cell-cultured meat and plant-based eggs. But it was perhaps especially apropos to ask it in this venue. Tencent is a huge deal in China—kind of like if you merged Twitter, Electronic Arts, Comcast, and Netflix into one future-forward juggernaut worth $500 billion. The people attending the conference were entrepreneurs interested in the next evolution of technology, and there was Tetrick, onstage with people demonstrating how to scramble his company's plant-based egg product for a live audience.

His speech lasted close to half an hour, and it landed well. It always does when he's abroad, far away from the United States. He told me as much in late 2018, when he recollected the feeling of being onstage in Beijing, wrapping up his presentation on JUST's work to remove animals from the agricultural equation.

"The energy in people saying 'I want samples!' was even higher," he explained. "You just have a rush of people to the stage as if I'm giving away diamonds. Whereas, if I'm in the US, you would have, like, people raising their hands and saying, 'Yeah, yeah. I'd have one. That'd be great.'"

This enthusiasm is fueling JUST's decision to increase its focus on international markets. So as Memphis Meats focuses much of its bandwidth on headway in the American regulatory arena (though Uma Valeti has said he believes the Asian market holds lots of promise), becoming the de facto leader of that effort, Tetrick has devoted an increasing amount of attention to foreign markets, particularly those in Asia.

While he does have to work and engage with food regulators in these places—including in Hong Kong, China, and Singapore—the environment is notably less politically charged, and unified around supporting food technology. He says it makes total sense to him to aim his attention toward these regions.

As I talked with leaders of cell-cultured meat companies about the temperature around the world with consumers and investors, it struck me how differently people think about food technology outside the United States. It struck them, too.

"There is an intensity around food in Asia that is different than the US," Tetrick said. "If I'm eating with someone that's in Beijing and I start explaining food demand and food scarcity and industrialized animal agriculture, in some ways they want to be like, 'Okay, shut up, I get this.' If I'm in New York, I have to explain it to someone a little bit more."

And it makes sense that for many people—investors in particular—outside the United States, the issue of food security is something that's all-too familiar.

"There's a more palpable sense of awareness of what's happening," Tetrick added.

It's a relief for Tetrick to deal with interest, rather than disdain, after his plan in the Netherlands collapsed, and after cell-cultured meat ran into regulatory roadblocks and industry skirmishes in the United States.

In 2019 alone, Tetrick spent about one third of his time in places such as Singapore, Hong Kong, Dubai, and China, underscoring just how important those markets have become to JUST. When the company released its plant-based eggless scrambled egg product in Hong Kong and Singapore, it caught on quickly. JUST expanded its operations to include twelve employees working out of Shanghai, all of them focused on laying the groundwork for getting his plant-based egg into restaurants and retail. And he sees that activity as key for the company's bigger future ambitions. He hopes to persuade Asian governments, especially Singapore's and Dubai's, to partner with the company long-term. It's not just about being the first place to serve cell-cultured meat to the public. Tetrick is also negotiating to find help with expanding his labs and manufacturing facilities.

"With Singapore and Dubai, it's not just a conversation about being the first place to serve this, it's about being the first place to build an infrastructure to feed this to the world," Tetrick explains. "Would they be open to investing in the labs and building larger cell manufacturing facilities?"

American plant-based burger company Impossible Foods is a guide. It's now a phenomenon, but it started small, in higher-end restaurants, and then slowly expanded as name recognition and cultural acceptance grew. Accordingly, Tetrick's immediate ambitions aren't modest—his goal isn't to produce tens of thousands of pounds a day. More like thousands of tons.

And while Tetrick acknowledges that, in many ways, much of the world looks to the US regulatory system as a gold standard from which to take a cue, he also feels torn about relying too much on the outcome in America, recognizing both the heightened interest and the huge markets outside his native country.

"I get the focus on the US, I don't get the obsession with it," he says. "I don't get it. We live in a big world and more people are eating meat outside the US than inside the US. I do think there is an overreliance on what the US is going to do. I wonder what

Singapore is going to do. What is China going to do? What is Israel going to do?"

Some regions of the world, such as Israel, are existentially threatened by the impacts of climate change, and are preparing for a world map redrawn by the looming crisis. That creates a lot of room for opportunity, though, too. Food security is of top concern in many of these markets. And it shouldn't be a surprise: The climate crisis is expected to create food access problems that will impact the lives of many millions of people.

Take Thailand as an example. In November 2019, researchers at Climate Central estimated that by 2050, the land on which more than 10 percent of citizens live is expected to be submerged. That could wind up affecting some 150 million people, including many who live in the metropolis of Bangkok, reshuffling how food is distributed.

In China, the World Economic Forum has said it expects increased flooding to threaten important spots for infrastructure, energy, and agriculture. China already has very little arable land relative to its population of 1.38 billion, so the ability to grow meat could one day be a major boon for the Chinese. Executives of cell-cultured start-ups in Silicon Valley, including JUST, who've worked with the Hong Kong–based investor firm Horizons Ventures say its top leadership, billionaires Li Ka-shing and Solina Chau, have an innate sense of duty anchored to the places in which they grew up and made their vast fortunes. Food security in China—the most populous country on Earth without enough arable land to sustainably feed its people—is expected to become a more dire concern as the world population balloons. Ensuring that they can be fed is a priority for Li and Chau, food tech executives say.

In March 2019, the National Research Foundation of Singapore announced it was going to spend some $535 million on the research and development of cell therapy manufacturing, including about $106 million for sustainable urban food production. The city-state has big ambitions to produce 30 percent of its own food by 2030—and that means there's space for an ambitious cell-cultured meat company to establish itself there first.

Tetrick says that this opportunity has changed the way he approaches his food products. If someone closely watched JUST over the course of 2019, they'd notice the company has been far more public about the expanding footprint of its plant-based egg product than its work in cell-cultured meats. This is on purpose, because Tetrick says his mung bean–based eggs are the key to his meat's success. JUST Egg suddenly represented a lot more than just another product to add to the company's suite of vegan foods.

"There was no way Apple could have nailed the iPhone without doing the computer first," he said.

Tetrick sees his plant-based egg as an early-stage computer, and cell-cultured meat as the ultimate future product. Plant-based foods are a runway for cell-cultured foods. Unlike any other start-ups in the business of making high-tech meat, JUST is the only one that is already an operational food company.

Tetrick has years of experience working with suppliers, retailers, food service companies, restaurant chains, and distributors—and those relationships are going to be invaluable as the company looks to introduce new food products to the market.

And this strategy—entering new markets with his plant-based products to help lay the groundwork for his future meat ambitions—plays to Tetrick's own strengths. He isn't an accomplished scientist or a notable policy wonk—but he can certainly weave a compelling narrative and sell food. It worked before when Li and Chau invested in the company early on, and it's happened again. While he went on a promotional tour for JUST's popular plant-based egg, Tetrick throughout 2019 also kept in near-constant contact with regulators in Hong Kong, the United Arab Emirates, and Singapore, all three of which are interested in being early adopters of cell-cultured meat.

By December 2019, Tetrick said he was all but sure that his meat would become available first in Singapore. He had dispatched JUST's own product engineers to the Asian market in previous weeks, packing up and shipping cell-cultured meat samples along with them to cook directly in front of a wide swath of high-level stakeholders, including a "large-format restaurant partner" that he declined to name at the time. The company had also played host to Singapore regulators, serving them its chicken products at its headquarters in San Francisco.

The effectiveness of establishing its plant-based egg to cell-cultured meat runway, as well established and smooth as Tetrick can engineer it, relies on one major unknown factor. Once cell-cultured meat does make it to market, popping up on some menu somewhere in the world, people will have to actually take to it. And that relies on a whole slate of factors that have to do with tough-to-predict aspects of our own nebulous inner selves— our cultural traditions and our psychologies.

13

THE TIES
THAT BIND

I n the early 1700s, a North African rabbi named Chayyim Ibn
Atar was making his way from Morocco to Jerusalem, which
was then under the control of the Ottoman Empire. While on
this journey, he worked his way up the west coast of Italy, pausing his travels to stay for some time in Livorno, a very small city
in Tuscany, not far from Florence.

While in Livorno, the rabbi came into some funding, which
he used to print his religious commentary on the Torah. Because
of his keen intellect and piety, those writings have remained of
interest to this day, primarily among circles of religious scholars.
But now, all these hundreds of years later, a handful of food
technologists in Silicon Valley are whispering about this obscure
historical figure anew, taken not by the rabbi's personal story,

but by something he had to say about Leviticus, the third book of the Torah and of the Christian Old Testament.

Chayyim Ibn Atar wrote that, in the future, God will literally alter the physiology of the pig so that it will chew its own cud and therefore be transformed into a state that is fit to be slaughtered, cooked, and eaten by Jews. It is not the Torah that will adapt to reality, he wrote, but reality that will adapt to the Torah. The laws of the Torah are immutable, but the nature of the pig is not.

It's an interesting bit of ancient history, especially for Eitan Fischer and David Bowman. Both men left JUST in 2018 to start their own company, Mission Barns, just a few miles north in Berkeley. Seeing an opening in the market, they immediately began working on the world's first cell-cultured bacon. And while this industrious pair has worked to culture their cells, some in the rabbinical community have been leafing through thousands of years of religious texts to figure out just what to make of cell-cultured meats and whether they would qualify as kosher. To be sure, not all religious scholars agree. But as their discussions have unfolded, some have been so bold as to suggest that this new food technology will ultimately render pork products and cheeseburgers admissible for Jewish palates.

One of those religious scholars is Yuval Cherlow, a Modern Orthodox rabbi in Israel who has openly talked about cell-cultured pork as being permissible to eat and still remain in line with kosher rules. Cherlow is one of six founders of Tzohar, a religious organization that seeks to find common ground and strengthen ties between religious and secular Jews, so it makes sense to me that he might take this position while more conservative members of the rabbinical community still blanch at it.

In an interview with the Israeli news service *YNet*, Cherlow explained that, in his opinion, if the "cell of a pig is used and its genetic material is utilized in the production of food, the cell in fact loses its original identity and therefore cannot be defined as forbidden for consumption." In other words, Cherlow believes that the process of cell-culturing a pig cell strips it of its inherent porky qualities, neutralizing it. The end product wouldn't even be meat. It'd be considered pareve, a neutral food that isn't classified as either dairy or meat.

Cherlow's opinion is not widely held within the rabbinical community, though there is precedent that lends his view some credence. About a decade ago, a powerful kosher certification group, the Orthodox Union, was confronted with a question about whether a common amino acid called L-cysteine could be considered kosher.

L-cysteine works as a reducing agent that relaxes bread dough by breaking down gluten protein in high-speed bread-making systems. Bakers use the amino acid as an ingredient to shorten mixing times, which results in less-stressed dough. That, in turn, improves the overall quality of the dough and allows it to be more easily divided and rounded for baking. It also can extend the shelf life of bread.

But there's a problem. Much of the L-cysteine that's used in bread baking comes from duck feathers, raising questions about

L-cysteine's kosher status because of rules surrounding how animals should be treated during and after slaughter.

Rabbi Menachem Genack is the CEO of the kosher division at the Orthodox Union, which is one of a small handful of Jewish groups tasked with certifying foods as kosher. If there are food products in your kitchen cupboard, there's a good chance they have been stamped with the tiny but common symbol of the letter "U" inside a letter "O." This is the mark of the Orthodox Union, and a signal that the food behind that label has been deemed kosher.

Genack ruled the process was kosher because, once the amino acid was collected from the feathers, it was run through a purification process before being crystallized, and that the process made its use in foods permissible. Cherlow makes a similar case for cells used to culture meat: that pork from the hair of a pig may be purified through its processing.

Genack is less certain. Though he has never made any attempt to muzzle his enthusiasm for cell-cultured meat in our conversations, he stops short of saying cell-cultured pork would be kosher. He lauds it as an innovation that is especially promising for the environment during a time in which climate change has understandably become a source of anxiety. In many religious traditions, being a good steward of the Earth, its inhabitants, and its resources has been described as our personal responsibility. His excitement for the technology is matched by the excitement of the people making cell-cultured meat, for whom religious dietary restrictions could be lucrative. The global kosher market is worth more than $24 billion; and the halal market, which includes food approved for Muslims, rings in at $1.6 trillion. Put simply, engaging with religious authorities to get their stamp of approval translates to big money.

"It's just such an exciting development," Genack has told me. "I'm amazed. The cattle industry and the meat industry, it would seem that they would be completely transformed."

The kosher question is just one piece of the puzzle. But it reminds us that how people use and interact with cell-cultured meat in our everyday lives will be the ultimate test.

I set out to report this book with every intention of trying to understand the process, result, and ramifications of cell-cultured meat from the ground up. I traveled to Silicon Valley, Israel, Amsterdam, and up and down America's eastern seaboard. I spoke to the world's experts on food technology, food policy, food regulation, and religious food law. But I quickly realized that I had to answer for myself a question that seems incredibly basic: What is meat?

A molecular scientist might think of meat on a chemical level—a collection of water, proteins, fats, and carbohydrates. Ask a farmer, restaurant chef, dietitian, or home cook the same question and the answer changes. For some of them, it's their livelihood; for others, it's a foundation of culinary artistry; and for still others it's simply a reliable and economical way to feed a family.

Meat is not purely a commodity, a raw good that is created, bought, sold, and consumed without accounting for the value it possesses by virtue of its history and place in human culture. As Michael Pollan pointed out in a November 2002 issue of *The New York Times Magazine*, humans have been eating meat for as long as we've been on Earth, and while we may not need it to survive, the ways in which we obtain, prepare, cook, and eat it have affected us tremendously in social, biological, and cultural contexts.

Throughout my own research I often posed these questions to friends over dinner, sometimes delighting in the inevitable pinwheel of contorted facial expressions as people worked through in their minds just how far they'd allow their culinary compasses to take them.

"Would you eat meat grown inside a beer vat?" I'd ask, bluntly. Because that's how many of the entrepreneurs in this food technology space characterize their ideal production plants for cell-cultured meat. They attempt to normalize the idea by conjuring images of little craft breweries, only instead of making beer they're growing bespoke meat. Some people within the food tech sphere have gone so far as to try to brand the stuff as "craft meat." Sketches of these production facilities can vary wildly. Some envision nondescript buildings that house rows of cylindrical bioreactors, standing tall like wide columns, in which animal cells happily proliferate until it's time for an earnest worker to come by and harvest the fat and muscle tissue for further processing. One group, a Tokyo-based nonprofit called the Shojinmeat Project, has released more stark futuristic renderings. In one imagining of a future meat growing operation in the year 2203, the people with Shojinmeat drew up plans for high-rise farm towers set outside urban centers—imposing skyscrapers dedicated to growing cell-cultured meat.

Interestingly, I estimate a good 85 percent of my dinner companions claimed they'd absolutely try cell-cultured meat; the other 15 percent were more inclined to purse their lips, aggrieved at an idea they often described as "just too weird." To those, I would say the same thing, that they should expect cell-cultured meat to become a reality whether they liked the idea or not, that there's too much money, science, and willpower working to push it into the market. Still, the disparity always struck me. If the vast majority of people with whom I spoke claimed to be interested— even eager—to try this new meat, that is surely good news for the handful of food technology companies striving to make it.

All along I knew there was something about cell-cultured meat that made me feel uncomfortable. On one hand, it's exciting to think of a food system that could eliminate a humongous source of animal suffering—more than 70 billion animals die each year around the globe to feed our omnivorous cravings. And during a time when ice caps are melting and sea levels are rising, how can we eschew an opportunity to shift toward a system that's theoretically much better for the Earth and the humans who live on it?

But how can we wholeheartedly trust these new food technology companies that promise solutions? Is this new food technology a clever fix, or is it an example of humans trying to create something too superior to nature, something that could very well have unforeseen and deleterious effects down the line? Is it genius, or hubris?

We have gotten really good at detaching ourselves from the natural world. Modern life allows a lot of us to ignore where our food comes from, passing our authority over what we eat to faceless entities—multinational food manufacturing giants and sprawling industrial farms that often operate to produce profit rather than nutrition and quality. In the developed world we can buy passion fruit in Wyoming without thinking at all about its long and complex journey from Brazil or Paraguay to a corner store down the street. This has contributed to a larger, more nebulous collapse in our value of place. We live in an illusory world in which twenty-four-hour delivery and easy access afford us privileged lives in which we conveniently avoid thinking about time, place, and the infrastructure that makes all these niceties possible and safe.

Big food companies have always cloaked their products and growth in the veil of health, modernity, and progress. But as we can see looking back through history at what those companies have wrought, their promises hardly even pan out. Just look to margarine, frozen pizzas, Big Macs, and deli meats; potato chips, soda, and every other now-ubiquitous food product that is packaged for our convenience and enveloped in a carnival of sugar, salt, fats, and a laundry list of unpronounceable ingredients. They are objectively bad for us.

But it would be unfair to judge cell-cultured meat based on the actions of previous industrial food giants. And there's no way out. Nearly every aspect of modern life is unnatural. We crisscross suburbia in gas-guzzling cars, seek respite in air-conditioned buildings, store data in digital clouds, and communicate with the world over invisible airwaves. Even the apples and bananas we buy in the grocery store are the product of human engineering.

I asked Princeton University ethical philosopher and author of *Animal Liberation* Peter Singer if he thinks humans are going too far, if it struck him as dangerous to take something as fundamental as meat and try to build it from the ground up in laboratories. Singer's 1975 work is the backbone of the modern animal rights movement, inspiring the founding members of People for the Ethical Treatment of Animals (PETA). His response was blunt.

"No, I don't, to be honest," he told me. "I think we can and always have striven to do better than nature. If we hadn't done better than nature with our food then we'd still be gathering grains and laboriously hunting for our meat. I don't think nature is in any way a gold standard."

Even if nature isn't, in Singer's terms, "a gold standard," the cell-cultured meat companies are doing their best to reach it. What is cell-cultured meat other than man's attempt to re-create something nature has already given us? But Singer is correct in pointing out that the history of humans is marked by our attempts to tailor the world around us for our own convenience, for better or for worse.

I'd heard this point of view before, but it had come from the mouth of the former chairman of Nestlé, the world's largest food manufacturer. On assignment for *Quartz*, I interviewed Peter Brabeck-Letmathe, who in his long career at the company positioned himself as the chief architect of a strategy that reenvisioned Nestlé not solely as a food company, but as a food company whose products might one day work more like medicine. In practical terms, that means fortifying foods with healthful nutrients that can prevent diseases—think frozen pizza that helps people avoid Alzheimer's disease.

Brabeck-Letmathe met me on a chilly December 2016 evening in New York at the St. Regis. The warm hotel lobby was draped in holiday decorations, mammoth wreaths, and long strands of forest green garland set against the hue of dim lighting. I was led up a flight of stairs by a Nestlé employee, who directed me into a fluorescent hotel conference room. A few moments later Brabeck-Letmathe swept into the room, sat across from me, and for the better part of an hour I listened as he explained to me his company's long-term ambitions. It struck me as creative and interesting, but it also unnerved me. I asked Brabeck-Letmathe about moving further from nature, and whether we should instead be turning back to the Earth for our foods. He shrugged off the idea.

"Your vision is basically saying that nature is good," he said. "That's wrong.

"Nature is not good to human beings," he continued. "Nature would kill human beings. The reason why *Homo sapiens* have become what we are is because we learned to overcome nature. You see, what makes us different from the animals, why we were able to develop afterwards, intelligently, was we learned how to cook. That is something that is completely underestimated."

And while Brabeck-Letmathe conceded that vegetables and grown foods play an important role, he told me that he knew it "wasn't enough" for him. For him and Singer, the agricultural revolution is an evolving narrative. The work of figuring out new ways to efficiently feed people is not work for the weary, though it doesn't make him popular with slow food advocates such as former *New York Times* food columnist Mark Bittman. I asked him about Brabeck-Letmathe's perspective and couldn't help but chuckle at his scornful reaction.

"It really is about, do you think shitty food boosted by nutrients is better for you than real food?" he said.

Similarly, when I asked Alice Waters, the chef of the iconic Chez Panisse restaurant and a leader of the locally grown, slow food movement, for her thoughts on cell-cultured meats, she recoiled.

"It just frightens me that scientists believe that they know more about it than Mother Nature does," Waters told me. "I really think of my food as deeply connected to nature. It has to do with seasonality, it has to do with a complexity of the soil that grows the vegetables that the animals eat. I think it's what nourishes us."

It's not that Waters doesn't believe in the power of science. She just isn't eager to disrupt the science that occurs when nature is left to its own devices. What we eat plays a major role in gut development and, as she put it to me, she isn't convinced people can do the same thing using test tubes.

Bittman and Waters may be right to be skeptical. Waters's worldview considers the lives of animals in far more gentle and respectful terms than that of the mammoth industrial meat and egg systems. They are an essential link in her hyperlocal, agrarian model, requiring our care and respect even as they ultimately exist, it seems, to serve our purposes.

"The animals of our lives that we can milk or we can get eggs from, we've always been close to and have cared for these animals in the past," Waters continued to tell me. "They are part of farming, their manure, they are part of this big picture of the natural world. I really need to get back to that."

In 1995 Waters established the Edible Schoolyard Project, which seeks to make the garden a classroom for students to learn about the relationship between seeds and soil, the growth of vegetables and fruits, and how it all fits in their world. The nonprofit functions on the premise that it's just not hard, through education, to teach children about the power of local food and how it can be used to support an entire community's health.

In so many ways, Waters and I are on the same philosophical page. I too am skeptical of consolidated agricultural companies such as Monsanto and Tyson Foods. The scale and pure mechanistic sprawl at which they operate has a numbing effect on my conscience. And I understand her lamentations about how, through the Industrial Revolution, we ultimately lost a very vital bond between ourselves, nature, and food—a bond that not only nourishes us, but also gives our lives meaning and a sense of context of where we sit in the fluid and evolving natural order of the world. "I'm sitting out in my vegetable garden right now and the beauty of it is just feeding me without my having to pick anything," she told me.

But Waters's vision for the food system, as much as I like the sound of it, isn't fit for a world that will be grappling with feeding a growing human population while also suffering from climate change. More droughts, bigger hurricanes, and increased flooding, to name a few effects, will reshape land and redistribute people and farmland around the globe. The warming planet is expected to recast the regions of the world that are best for growing food. A 2018 report by the United Nations Food and Agriculture Organization, for instance, said food yields in certain places are expected to drop by close to 3 percent, including in West Africa and India. Meanwhile, Canada and Russia will prove to be some of the more optimal spots for growing grains.

A lot of that might seem far away from home and thus easier to ignore, but Waters's own California is expecting to face a host of changes that will make growing food a tougher business, to say nothing of the hurdles of trying to shift communities into being more responsible for growing their own food. The state says it anticipates more frequent and severe droughts, less water stored in the Sierra snowpack, an increase in the number of pests and invasive species wreaking havoc on crops, more heat waves, and fewer chill hours, which are imperative for farmers growing fruits. The California Department of Food and Agriculture is currently working with farmers, ranchers, and food processors to educate them about the risks of what's coming, and to brainstorm and develop potential strategies for withstanding the problems of the future.

Waters's plan might go over well in some very select communities, but there's not enough time and there are too many challenges barreling toward us to try to switch something as monolithic as the meat system away from its industrial scale and down to locally raised cattle, chickens, and pigs. The solution is somewhere in between, and I believe it likely will be found in places like the laboratories of JUST.

Which is not to say that we shouldn't be wary. These young, high-minded start-ups might be more idealistic, but even if they one day deliver a blow that puts industrial farming out of business, we would be trading one secretive food entity for another. After all, whether you trust Tyson Foods to make your meat or JUST, people are still relying on a mostly secretive and mechanistic system, estranging people from the humanity of food production. It's a far cry from Alice Waters's community garden.

Industrial farms today are just as opaque—if not more so—as they were then. To an alarming degree they use medically important antibiotics for preventative and treatment purposes; many of these businesses support so-called ag-gag laws that forbid filming and photographing of farm activity without the owner's consent; and they condone poor treatment by workers of animals. By the same token, companies behind cell-cultured meat are only as good as their closely guarded intellectual property. Cell-cultured meat marketing campaigns have produced polished videos of spry young chickens, freed from the tyranny of slaughter, clucking freely across lush settings, accompanied by hopeful footage of scientists peering into microscopes. All the while a narrator explains inspirational sweet-nothings about all the good that was marshaled to make this product a reality.

But the reality is we simply don't know *exactly* what's in JUST's liquid medium. Behind the white lab coats and sterile laboratory walls is a process to which people will never have unfettered access. And even if we did have a deep insight into the processes these companies employ, the average person would need a graduate degree in science to fully comprehend their methods. While this isn't a deal breaker—after all, I don't feel the need to be an electrician to enjoy a house with lighting—it is important to remember this technology isn't simply a fancy invention for industry disruption. It's food. How much should we be able to wrap our heads around how something is made before we eat it? At the very least, we should expect and demand the maximum amount of transparency.

The entrepreneurs, scientists, and activists pushing the envelope of food technology so often spoke about this invention as just a piece of technology, at times seemingly blind to its meaning, or even its inherent sacredness. It is an astounding advance, perfectly engineered to disrupt the meat industry and work wonders for the environment along the way, but deep in my gut, this rhetoric makes me uncomfortable. This stood out like a sore thumb during an August 2018 discussion with Pat Brown, the CEO of Impossible Foods. To be sure, his company is not making cell-cultured meat. Brown sells relatively convincing plant-based burgers, but his view of food as technology perfectly encapsulates everything that makes me nervous about how so many people in Silicon Valley's cell-cultured meat space approach their food inventions.

During my conversation with Brown, he mentioned that his adult children had never eaten meat during their whole lives until his creation hit the market. I paused.

"You would label this as meat?" I asked, pushing at the sides of an Impossible Burger that had been set before me. Its main ingredients are coconut fat, wheat, potato protein, and genetically engineered yeast to produce its heme, an iron-containing molecule. It contains no animal cells, conventionally sourced or cell-cultured.

"Absolutely," Brown replied. "I mean, animals have just been the technology we have used up until now to produce meat, which is a food that is defined by its flavor profile, its sensory profile, its nutrition, utility, and stuff like that. What consumers value about meat has nothing to do with how it's made. They just live with the fact that it's made from animals. If we're producing a product that is delivering everything that is of value in meat for consumers, it's filling that niche. It's the right label for it."

I sat silent for a moment.

"So it's ultimately not a molecular question?" I asked. "It's not what you see under a microscope, it's about what people experience?"

"It's what functional role it plays," he said.

It should be said: Brown and other makers of plant-based meats might tout their products as analogous, but they still don't match the exact taste or even mouthfeel of real meat. Not yet, anyway. The companies might offer basic burgers and sausages, but those products represent an infinitesimal speck in the broader constellation of meat dishes that have been prepared around the world for centuries. Meat is more complicated than high-tech engineering and mimicking flavor and texture experiences with fancy industrial gadgets. And perhaps more important is that the processes by which these new products are made are still shrouded in industrial mystery. Replacing the feedlot cow that brought you a cheeseburger with meat grown under top-secret conditions is still, at the end of the day, an alienating food system.

Whether we have the money to eat the fanciest caviar or only enough for a humble potato, we ingest food the same way. We are all equals when it comes to satiating our hunger, and for that reason our relationship with food is highly intimate. It becomes even more so if your family has passed down barbecue recipes for generations or *mocotó* from Brazil, *pašticada* from Croatia, or even my mother's own pot roast recipe, which simmers away for hours. As cell-cultured meat companies gear up to introduce their products to the public in restaurants, food service, and grocery stores, important questions will emerge about how these cuts of environmentally friendly, man-made meats will impact the culture of cuisine. Are they kosher and halal? Will a generations-old barbecue recipe be possible with cell-cultured meat? How about Julia Child's famed beef bourguignon? These are important questions for a young industry that's aggressively seeking to lure committed meat eaters.

These dishes have special cultural significance, and the industry seeking to reshape the meat system as we have known it should heed the importance of those cultural touch points—its success may rely upon them.

It's also sad to think about a food through Brown's lens, solely as the sum of its functions. It's just less interesting, plain and simple. As food historian and author Sarah Lohman once put it to me, "I feel very bored when I hear about any food being reduced down to a functionality. A great comparison is that sex is just a way to make babies."

Still, as I weighed for myself the benefits and downsides of cell-cultured meat, it was incredibly difficult to come up with a reason to reject it. I am cautious of it and the companies. Overcoming the technological and scientific challenges to make this meat a reality has required a herculean effort. Navigating the regulatory minefields around the world has also been tough. But maybe, just maybe, the biggest challenge for cell-cultured meat companies will ultimately involve how it fits into a culture that already has a long and rich relationship with meat. If scientists can truly create a meat replica that honors and respects the types of cuts and traditions we've incorporated into our diets and recipes throughout human history, it would seem a great waste not to embrace the opportunity to take advantage of the environmental benefits.

When I began thinking more deeply about cell-cultured meat, I wondered what my grandmother would think of it. I still remember racing through her cozy kitchen as a little kid, jostling against my cousins in a rambunctious pack, tearing through the tidy rooms of her house only to be ushered back outside by the adults. Even in those moments of chaos, the aroma of food was unavoidable and distinct. It seemed like the stove was always on in her home, something was always simmering or frying. And a lot of those dishes incorporated meat.

Her initial reaction to the idea of cell-cultured meat was to bypass the cell-cultured aspect altogether. In fact, to my surprise, she was eager to talk about how she felt the smell of conventional ground beef has changed over the years.

"It smells like chemicals," she told me, assured that the modern industrial meat manufacturing apparatus has gone too far in its processing of our food. She explained how she'd recently opted to bypass beef for bison. It smelled more like meat to her and tasted good, too.

That she was open to changing her meat consumption habits signaled to me that there might, under the right circumstances, be an opportunity for cell-cultured meats to make a good first impression.

I returned to her house in rural Raywick, Kentucky, months later. She was cooking dinner alongside her youngest daughter, my aunt Paula, while I updated them on my reporting.

"So would you buy it?" I asked.

She was grating cheese at her kitchen counter. My aunt stirred a bubbling red sauce with a wooden spoon.

"I'd have to get past the whole thought of it all," my aunt replied. "So what does it look like?"

I described the foie gras I'd tried at JUST, the chicken tenders at Memphis Meats, and the thin steaks being produced at Aleph Farms. My grandmother turned around and leaned back against the counter, brow furrowed, and slowly shook her head, unconvinced.

"I think I'd rather have the real thing," she said, turning back to her grating.

"But you may try it and like it," my aunt said. "Who knows?"

"It's probably too expensive anyway," my grandmother replied.

"What if it's cheaper or the same price?" I asked.

She paused and nodded her head to one side as she considered the possibility.

"I guess if it tasted okay, yeah."

The reluctance in her general response to my question was expected. What I didn't anticipate was her apparent willingness to give cell-cultured meat a shot, provided it looked, tasted, and cost the same as the meats she'd grown accustomed to cooking and eating. That's most of the battle, I feel, when it comes to this new meat. If people are willing to try it—if my own grand-mother is willing to try it—then cell-cultured meat companies probably have the potential to establish a fairly wide audience of consumers who'd seriously consider adopting it into their diets. But cooks and consumers need to be able to handle and cook it just as they handle and cook conventional meat. Easier said than done.

I glance over at my aunt's two sons, a couple of my youngest cousins, and I wonder what kinds of foods they might be eating in the next decade or so. I wonder what recipes will be passed down to them, whether the concept of cell-cultured meat will give them pause, or if someday it will just be an accepted product to choose from at the grocery store. I wonder what the future planet will look like for them, too.

During the course of my reporting, my attitudes about meat changed, almost necessarily. As for many people, food, and especially meat, is part of my cultural heritage. What happens when heritage collides with ethics? Aside from the proposition posed by cell-cultured meat companies, I've also come face-to-face with questions about my personal responsibility to operate as a good steward of the planet. And food—eating meat—is one of those odd areas of life where culture and climate collide at the dinner table, and it has forced me to take stock of the way I live and interact in the world. The questions I confronted hit a very personal chord.

It wasn't until my early twenties that I realized just how powerful the art of cooking could be for me. Shortly after coming out of the closet, I experienced what a lot of queer people go through when they begin to maneuver through the minefield of emotional and psychological self-acceptance. And when you're learning to love yourself, there's a natural desire—a need, even—to surround yourself with a supportive group of people with shared experience, often described by the LGBTQ+ community as "choosing your own family." And it can be essential for queer people, particularly those who are vulnerable to the many hostilities of a world that, even today, far too often recoils at their existence. This act of self-preservation unfolds differently for everyone. Making a home for yourself is a truly singular experience.

For me, finding this community involved putting food and cooking at the center of my world. Cooking became an outlet through which I was able to tell important stories and express myself. It's a sentimental way to think about food, but I do believe that elements of who we are and where we come from are revealed in the aromatic swirls drifting from a kitchen and the lingering tastes in our mouths. Our relationship with food, what we cook, and how we cook connects us to prior generations and colors our present. In the art of cooking, food is the paint we use to create experiences. It helped me create a home.

While living in Washington, DC, I began to regularly host small dinner parties, in which the people seated around the table were almost exclusively queer.

The rules were simple: No more than eight people were invited to any given dinner, everyone had to be meeting at least one new person (including the host), seats were assigned, I did all the

cooking, and guests brought wine. Each meal was the culmi-
nation of careful planning, inevitable mishaps, and the art of
improvisational flourish. My dining room table became a safe
space, an equalizer in a sliver of time and place at which we
could collectively let our guards down and share ourselves. Cooking became a creative outlet, a form of expression that helped me
claim confidence, and a way to showcase levels of exuberance. It
was a way to nurture.

Eventually I moved to New York, where everything down-
sized out of necessity. The move required getting rid of a dining
table. The dinner parties didn't stop, though. Once I regained
my footing, I began hosting them in my new home, cramming
new friends around a folding table, which was happily weighed
down with homemade dishes, wine, candles, and cutlery.

And at first meat played an important role. I attempted beef
bourguignon, labored over grilled chicken marinated for hours,
and experimented with lamb. I borrowed a page from my nana's
own recipe collection, cooking fresh green beans in boiling water with bacon.

But as I began to pay closer attention to what scientists were
saying about the climate crisis, and as I reported more deeply on
the work of cell-cultured meat companies, it became clear to me
that, in order to be a better steward of the planet in my own
small way, I would have to change my habits.

We must acknowledge the agricultural reality outside our
kitchens. It is impossible today to be a responsible and conscientious consumer of food without also heeding humans' own role

in the story of our changing climate. Since animal agriculture accounts for at least 14 percent of total greenhouse gas emissions, every cook has a set of decisions to make about what they feed themselves and their loved ones. As a result, I attempted to navigate a new landscape of ethical questions around buying and serving meat.

It isn't easy. And for a while my cooking tapered off.

Of course, the obvious path forward was to try to reinvent my approach to meals. I know preparing exclusively plant-based dishes can be exciting. But still, it was hard for me. Like so many humans, I'm part of a larger story that has placed meat at the center of the plate, the principal actor on a stage shared by a supporting cast of fruits and vegetables.

How, if at all, does removing meat change the nature of a meal? Is it ethical to serve meat? Would I have the appetite to host a dinner in which all the comforting scents and tastes of meat are absent, replaced instead with a military parade of broccoli, squash, eggplant, and sweet potato?

Many of us, from Rabbi Cherlow to my grandmother and me, face these questions. How to reconcile the past and the future on our plates? It's okay to be stumped by some of these big questions, and it's okay to make small changes. I stopped eating beef and pork, limited my chicken intake, and ate more responsibly farmed fish, such as salmon. I expect my habits will evolve more as time passes. For other people, it's about giving up meat once or twice a week, or sometimes even more. The decisions we ultimately make about what we should cook and eat are personal.

We all have different values. But it is exciting to think that one day all of those concerns might vanish because of a bold new technological advance that allows us to connect to our pasts without dooming the future. Certainly I look forward to a day when I might be able to ethically make my mom's pot roast without the guilt of knowing it came from an unsustainable segment of the larger food system.

In 1846, Abraham Pineo Gesner invented kerosene, saving the lives of countless whales, which had been hunted for their oils that lit streetlamps. In 1908, Henry Ford unveiled a car that would make traveling by horse obsolete. Neither of those men were animal rights activists or had an inkling about climate change. They lived very different realities from each other and certainly from us, and yet their work has played an undeniably outsized role in advancing the interests of people plunging into the next phase of modernity. The promise of cell-cultured meat—whether it's first brought to fruition by Josh Tetrick, Uma Valeti, Mark Post, or some other intrepid entrepreneur—leaves me feeling optimistic about what's next, even if it is speckled with all sorts of interesting and unanswered questions.

Almost three centuries ago Rabbi Chayyim Ibn Atar pondered the fundamental nature of our food: what we can change, and what we can't. Now the rise of cell-cultured meat asks us to pose a similar question about the role of meat in our lives. Will our desire for meat have to change because of reality, or will reality have to change because of our desire to eat meat? Can we switch mostly away from meat to support a healthier, more sustainable planet, or do we reinvent how we approach meat in a separate bid to accomplish the same thing?

One thing I am certain of: We are in a unique and important moment in food. The changes in the food system we're seeing, and indeed the entire cell-cultured meat movement, are fueled by passionate people who have seized a specific moment to drive the change they want to see. That's noteworthy and singular. And it's up to the rest of us to be thoughtful decision makers, to consider those products with a brighter view of the future in mind and embrace our own roles in helping to usher that vision forward.

SETTING
THE TABLE

'm nervous," she tells me.

We've reached the top of a steep set of stairs and I turn to my mother and give a supportive smile.

"Don't worry about it, truly," I say, a fruitless attempt to reassure her.

But my mom, Connie, has been plucked from her regular day-to-day life in Louisville, Kentucky, where she and my father raised me and my three siblings. Of course she's feeling nervous. It's not every day you find yourself stepping into a Silicon Valley food technology company, where you're expected to talk with a brash young CEO and walk through sterile laboratories.

We step forward into a spartan waiting area where I sink into a sofa. My mother walks up to a large window and gazes out onto Folsom Street. It's damp and chilly outside.

Andrew Noyes, JUST's spokesman, appears from around a corner and glides into the room.

"Good morning! Welcome to JUST," he says to us, extending a hand toward my mother. "I'm glad you could make it."

She pulls her purse strap snug over her shoulder and takes his hand, smiling.

We had arrived in San Francisco the previous evening, after taking separate cross-country flights—mine from New York, hers from Louisville. It had been nearly two years since I embarked on an effort to learn about cell-cultured meat, about two years since I'd first stepped foot into JUST to try the company's cell-cultured foie gras and duck chorizo tacos. Now, on the Tuesday before Thanksgiving in 2019, Tetrick was going to host a lunch and my mother was invited to join—a representative of JUST's potential consumers.

The truth is, I am nervous, too.

No one person in my life has done more to inform my approach to food than my mom. Long before any of Michael Pollan's books, she already had me gravitating toward fruits and vegetables and away from sugar- and salt-laden packaged goods. Meat eaten in moderation was key, and a cautiousness about where we get and how we handle meat was paramount. It's because of her that I linger in my grocery store's aisles, carefully examining the labels on my food. She's also the reason I've come to expect full transparency from the companies that want to feed me and the rest of America.

I'd talked with her some about cell-cultured meat, as I reported and wrote stories about it, and she remained mostly skeptical. Its perceived proximity—or lack thereof—to nature was alarming to her. Eliminating a living, breathing animal from the equation was a concept more likely to raise an eyebrow than curiosity.

My mother is precisely the kind of person Tetrick needs to successfully persuade to try his food if he expects to attract a lot of people to cultured meat. She's the on-the-go mother of four who generally shops for food at chains such as Kroger or Whole Foods Market, carefully selecting meat, balancing quality and price. She's especially conscientious about the quality of life the animal lived before it was slaughtered, and she pays extra-close attention to how she handles the meat once she's back at home, leery of pathogens that can cause food poisoning.

My mom's attitudes about food are generally closer to Alice Waters's than Tetrick's. When she shops for produce she looks for fruits and vegetables grown according to organic standards meant to promote ecological balance and biodiversity conservation within the food system. When I was growing up, my parents' skepticism toward the meat industry was a major factor in their decision to purchase two black Angus heifers, breed them, and raise the resulting steers for food. I was too young to understand the merits of this plan at the time, other than the irritation of adding a few extra chores to my list, but in hindsight it makes perfect sense. My family wanted to know exactly what the cows were being fed, so we reclaimed our authority within the food system and took control of what we were eating. In doing so, we eliminated many of the transparency concerns created by the meat industry. Obviously not everyone is in a position to do this; my family was privileged enough to have the option. After a few years, we switched back to buying meat at the grocery store. Raising and keeping cattle isn't easy work. Curious and mischievous, the cows would sometimes find ways to wander out of their fenced-in field, sparking frenzied efforts to chase them back to where they belonged. And it was tough work corralling them into trailers that we had to borrow for transporting them to breeders and to a slaughtering facility. In theory, though, people should always have the ability to step back and reclaim the authority we all delegate to food companies that produce dietary staples. Ours was a substantial investment, but one my parents do not regret.

My mom has been so committed to learning about food on a close and focused level that, after more than two decades working as a cosmetologist, she decided to go back to school to study nutrition, carving out time in her evenings to pore over the complexities of biochemistry, including the functions of macronutrients, micronutrients, and the ways in which they interact with our bodies. She works in a public school cafeteria now, directly handling foods that are being fed to children each day between classes—experiencing up close how the food system operates at scale.

None of her choices were easy. Buying organic can often be up to 30 percent more expensive than buying conventionally grown food. Raising cows is a really tough endeavor that requires a whole new level of commitment to one's nutritional beliefs. And certainly going back to school to study a tough field of science shows moxie, mettle, and a profound determination to figure out and detangle the best way to make ethical choices within a system that doesn't always leave much room to maneuver.

If Tetrick can't persuade her to think twice about cultured meat, I would be concerned about how eventual consumers would react to his product.

Before lunch, Noyes walks us through a tour of JUST's headquarters, taking us down into the plant library that I saw so many months before. The room is packed along its walls with floor-to-ceiling shelving units, weighed down with plant samples from around the world that have been mined by JUST scientists for their protein qualities. My mom steps toward the center of the storage room where, on a display table, a bowl of mung beans sits. This is the plant that has become crucial for the company, as it is the main ingredient for its vegan liquid egg product.

My mom and I continue our tour through the first floor of JUST, following Noyes around a corner and toward another set of stairs. We pass an area of the company I've never seen before: JUST's newest cell-cultured meat laboratory, built in the spring of 2019. Behind its large glass doors, two scientists in white lab coats and safety glasses are milling around inside, maneuvering from desktop computers, microscopes, and a large temperature-controlled storage device where small flasks of liquid medium are being automatically stirred, the little cells inside replicating over and over again until they soon will grow into small samples of meat. This lab is dedicated to research and development for eventual commercial use. The original lab, upstairs, is now used mostly to grow meat for the company's on-site test kitchen, where product engineers figure out how to best cook and prepare the meat.

My mom and I are led up a flight of stairs, down a hallway, and into the large open floor plan work space that is the nerve center of JUST. Noyes leads us past rows of people sitting at their desks, to the center of the room where we sit with Vítor Espírito Santo, the director of the company's cellular agriculture division. He speaks with my mom about the basics of the science behind what goes into growing meat from cells, the biggest challenges for the industry, and the progress the company has made in recent months.

I look over Noyes's shoulder and over toward JUST's product development kitchen. Chef Nate Park is busying himself at one of a half dozen workstations. Park used to work at Moto in Chicago, a Michelin-star restaurant known for its molecular gastronomy style and high-tech dishes. That restaurant drew attention from the trend-conscious, culinary-curious crowd, who eagerly sampled strange menu items such as carbonated fruit and edible paper.

Earlier in the morning I'd received a text message from Noyes with a photo of Park's handwritten menu for lunch. First up was a grilled chicken salad appetizer followed by a fried chicken entrée. Tetrick materialized from behind a doorway nearby, signaling we were ready to sit down and eat.

I look to my mom and smile. Her posture is more relaxed than before the tour and sit-down with Espírito Santo. She's intrigued by what she's seen and learned about the process and even more curious to see the resulting meat.

"Ready?" I ask.

Tetrick greets her and the three of us step to the kitchen. A rectangular table with three stools lined up along one side beckons us.

As we take our seats at the table, Tetrick launches into a part of his personal narrative that I've heard before. A folksy way for him to connect the real world outside of JUST to the work that he's doing inside the company.

"I grew up in the South, in Alabama," Tetrick says, sharing that some of his most palpable memories were of ordering oversized, deep-fried chicken strips that were served up alongside honey mustard and handed over in big clamshell Styrofoam containers. "I grew up eating as much meat as anyone, and I believe we have to figure out a way to give people the meat that they want without all the other things that unfortunately come along with it."

"Right," my mom says. She clutches her purse, which sits firmly in her lap. We are both still attempting to ease into the experience we're about to have. Tetrick continues.

"There's a way to give some boy from Alabama his chicken strips without causing all the harm by eating them," he says.

Park steps up to the table with his first plate of food.

The dish contains eight small circular discs of brioche topped with grilled chicken salad. The chicken has been chopped into small cubes, mixed with celery, pecans, dill, pepper, salt, and JUST's vegan mayonnaise. I reach down and pick one up, eye it closely and take a bite. The texture of the chicken is exactly what I'd hope for. The taste, though dominated by its seasonings, holds up to my expectations as well. It tastes like my mother's chicken salad.

Next to me, my mother takes a bite and almost immediately signals her approval of the appearance and taste. She chews for a moment, while listening to Tetrick as he describes his goal of making this meat better and cheaper than the conventional kind she can find in any grocery store.

"What do you think of this?" he asks her.

"I think it's delicious," she responds. The enthusiasm in her voice catches me off guard. "Can I get this recipe?"

"We need your recipe!" Tetrick says.

"I mean, it tastes like meat," she says. "It tastes like chicken— and that's what it is."

Nate Park comes over from his workstation, this time bringing over a plate of freshly baked biscuits and some vegan honey butter—a precursor to the fried chicken he's preparing. He looks down at our plates, seeing them empty, save for a few stray crumbs.

"What did you guys think of this?" he asks.

"Oh it's awesome, man," Tetrick says. "Just awesome!"

"Very good," my mom replied. "What did you think of it?"

He shifts his weight to one of his hips and brings his hand to his chin, rubbing it.

"I've made a lot of chicken salad in my life, but I've never made a cultured chicken salad," Park says. "Grilling it is a relatively new phenomenon."

When he first started working with JUST's meats, Park mostly focused on processed meat products like chicken nuggets that are easier to make. But then he said he began traveling more to China and Singapore as Tetrick began making new inroads into those markets, pitching the concept to interested parties across Asia. One of the first questions those people would ask, he says, was "Does everything have to be deep fried?"

"No," he says, smiling at us. "It doesn't."

Curious about the cooking process for the food she just ate, my mom asks Park how he went about preparing the chicken. Can it withstand being marinated? Not just yet. Are there bones? No. Did he put anything on the grill, she wonders, any sprays?

"Just a little canola oil, just like I would normally," he says. "You treat this product basically the same as you would anything else. Once it's done it looks like chicken, it acts like chicken, you cook it the same way."

She nods, still curious about how the food goes from lab to grill to plate. In this case, Park took a frozen package of thick, pinkish-beige paste—it looks a lot like a chicken breast—that was originally produced using the company's own in-house liquid medium (which now costs between $1 and $5 a liter, according to the company) and a 1,000-liter bioreactor. The cells were mixed with mung bean isolate to give the product a more defined shape. The final slab of chicken was 75 percent cultured meat and 25 percent mung bean.

Tetrick describes the project in technological terms. Just as there have been different versions of the iPhone, for instance, there are different versions of the company's cultured meat. This is "generation one" cultured meat, he says.

Park returns to the kitchen to finish preparing his main course. At this point, my mom has become an impromptu focus group for Tetrick. And of course she has. Day in and day out he interacts with meat industry leaders and people who devote huge amounts of bandwidth to thinking about the future of meat. Now he's sitting next to someone who doesn't have a stake in his success. He wonders aloud what would be the main factor that would keep people in Louisville, Kentucky, from trying cell-cultured meat.

My mom thinks for a minute, clearly somewhat uneasy at the idea of being asked to speak for an entire city. But then she says exactly what I'd imagined she would say—and it's exactly what I would have said.

"Because a human made it," she replies. "That would probably be it."

"Human-made chicken," Tetrick muses, then starts to chuckle.

"Sorry," my mom says. "You know, it's like . . ."

She pauses because she's in a tough spot. It's not easy to tell someone to their face that an idea they've spent millions of dollars trying to perfect is a tough one to sell simply because of a narrative hang-up. She believes JUST's cultured meat is actual meat on a molecular level. She's seen it, touched it, and now she's tasted it. But how do you overcome a mental barrier? Seeing that he's made her a little uneasy, Tetrick quickly admits that he sees exactly where she's coming from in her reply.

"I wouldn't want to eat human-made meat," he says.

At some point, he says, the companies behind cell-cultured meat will have products that taste exactly like conventional meat. The texture will be on point, too. And eventually the price will be competitive. But, he says, it'll be a tougher problem trying to convince people who have a strong and immovable preconceived notion that meat has to come from killing an animal.

The table goes silent for a moment.

"I feel more wobbly on that one," he says. "You can have the perfect technology, the perfect taste, the perfect everything, and at the end of it all, if folks in Louisville think this is a human-made chicken, it's not going to work."

To my surprise, my mom tries to reassure him.

"It's just the newness of it," she says.

She invokes the rise of the Impossible Foods burger and the widespread availability of Beyond Meat products. Those didn't take off immediately, either, she points out. But now they are popular, and people eagerly look to try them.

"This is going to have to go through some of those things too, I think," she says.

Park approaches the table again, this time with dishes that waft the familiar smell of fried chicken.

Arranged on slate-gray plates are thin cuts of breaded and fried chicken. They sit on a basic sweet potato puree that's been sprinkled with roasted cauliflower, brussels sprouts, and a little zucchini. It looks more like a gourmet meal than any sort of typical southern fare. Tetrick tells us that the meat on each plate is worth about $100.

I see the meat and immediately think back to the duck tender I tried at Memphis Meats earlier in the year, how humbled and impressed I was at how far that company had come in making a convincing piece of meat, with all the texture of the chicken I had always been accustomed to eating. What's in front of me is not quite close to that experience. The taste is there, but the texture of the meat is still, in my own opinion, more akin to a thick frittata than to fried chicken.

"What do you think when you cut into this, through the breading?" I ask my mom.

"I think it looks like chicken," she says.

"And the taste?"

"It tastes like chicken," she says. "It's a little bit different from chicken but it's really close. It's a good difference, you know? It tastes like a high-end chicken patty."

She turns to her right and asks Tetrick what he thinks about the dish but he can't answer her. He admits he's still stuck on her characterization of cultured meat as being "human-made meat." Maybe the issue, he muses, is that when cultured meat is being talked about, it's often through the concept of the bioreactor—the place where the cell culturing actually happens.

What if the narrative was instead linked more to cell collection? he wonders.

He mentions that the company sources its beef cells from three places: cows in Patagonia, an Angus beef farm in Marin County, and a Wagyu beef farm in Japan. What if people in Louisville associated the cell-cultured meat in their future grocery stores with the farms where the cells were sourced. Would that help ease some of the yuck factor?

She nods. It would, she says. He continues his line of thought.

"You are breaking a paradigm that has been so part of the culture—it's so nostalgic and traditional," he says. "How do you introduce that?"

It's a question that naturally bleeds into our next subject of conversation, which is that of naming and labeling. Can you simply call what we're eating chicken? Tetrick wonders. He asks me to reach and grab a nearby bottle of the company's JUST Egg liquid scramble product. He holds it up for my mother and points to the center of the bottle, where the label simply says "JUST Egg," then points to the bottom left-hand corner of the bottle where in smaller text "plant-based scramble" is printed.

"I think, to be fair, I would like to know if I was buying cultured chicken or conventional chicken," my mom says.

"Is it fair to call it chicken, though?" I ask.

"I think so because its base is chicken cells," she replies. "It's chicken."

And then something surprised me.

Tetrick took his finger and pointed to text on the bottom

right-hand side of the bottle, text that was much too small for me to read from where I sat a few feet away, but mentioned that the product was cholesterol free and non-GMO and stated the ounce size of the bottle.

"My preference would be to make it as small as possible," he says, tapping the bottle where the tiniest text was printed.

My mom leaned back, seeming surprised.

"Aren't you proud of it?" she asks. "Why would you want to hide it?"

It's a good question—but it's also one that reveals something about Tetrick. In a turn of events that I never would have anticipated, it was as though, in that moment, my mother was more assured about the future of cell-cultured meat than the man in front of her. Whereas Tetrick sought to minimize what the food was, my mom was inclined to pump up that aspect. And she wasn't suggesting as much because she believes JUST needs to err on the side of transparency. She thinks the product is neat.

The dishes before us are being cleared as my mom takes the last bite of her lunch. She folds her napkin and we stand up to collect ourselves. I turn to Tetrick, who is already being beckoned by JUST staffers to tend to a group of suited visitors who arrived while we were eating.

My mom shakes his hand and offers a supportive final thought.

"It tasted perfectly, and you know, I didn't expect that it would," she tells him. "In my head, I knew it was meat, it's just in a different form. But the taste was good, and I think it's probably going to go over well when you get it going."

He grins and thanks her, then turns to me and shakes my hand before whirling around and jogging to another corner of the room where the group of people are waiting.

My mom buttons her jacket and looks up at me. We did it.

Noyes eventually leads us back down a long hallway from the office space, down a flight of stairs and back to the entrance of the building, where we emerge onto Folsom Street again. It's wet from rain. We slide into a cab and I wonder, briefly, what we'll have for dinner later. I wonder how much the two of us might mentally grapple with the meat on our plates, knowing there's another way to produce it.

We drive by a stretch of the Mission District that's heavy with colorful hole-in-the-wall taquerías. Pink and pastel-green storefronts are framed by inviting signs and meat-laden menus taped to their windows. Little do these shop owners know that a man just a few blocks away is pouring millions of dollars into a promise that could fundamentally challenge their future generations from taking on some of these same recipes. The makers of cell-cultured meat have a lot to get right.

My mind returns to those shops when I make my own versions of these dishes or smell them wafting from the apartments of my neighbors. I think about the hours spent slow-cooking chicken or beef, completely powerless against those irresistible scents as they drift into the stairwell of my postwar apartment building back home. Even a few floors down you can catch whispers of these aromatic incantations, stopping any food lover in their tracks. Re-creating the meats used in those culinary traditions will not be easy.

The biggest barrier for cell-cultured meat may be the many people who have yet to try it, who remain skeptical or unaware of this technology, or unsure whether it can fit into their lives and traditions. Tetrick often talks about making an effort to reach out and appeal to folks in Alabama, where he was born, rather than those who've made San Francisco, Los Angeles, or New York their homes. But there aren't two versions of America—one of coastal elites and one of so-called regular people. The fact is that no matter where you live in the United States—or around the globe—the future of meat touches us all. At the end of the day, we all experience meat in the same, rudimentary way: We chew it, taste it, swallow it, and digest it. We will all have the same questions about how cell-cultured meat stacks up against conventional meat.

I'm heartened by my mom's experience. She walked into JUST feeling incredibly skeptical. She did so with an open mind, though, and by the end of her time there she was able to walk out imagining herself buying a package once it someday reaches the Whole Foods Market or Kroger where she does her regular grocery shopping.

But as Tetrick points out to us, not many people will be afforded the same opportunity to visit JUST's headquarters. Not everyone will be able to personally sit down with him, or Mark

Post, or Uma Valeti, or Mike Selden, or Shulamit Levenberg. How do you prepare people for going into their local grocery stores and picking up a package of cell-cultured meat and not being revolted by it, and by the idea that it's "human-made meat"?

I've wondered over and over to myself if, with so much at stake, Josh Tetrick is the right person for this task. He publicly talks about wanting to be the first person to get this meat to consumers—and if he succeeds it's the meat that JUST produces that will be the world's first example of how analogous cell-cultured meat can actually be to the conventional kind. I know there are companies out there with more impressive science behind their meats—I've visited them. I know there are CEOs of cell-cultured meat companies who have just as much drive as Tetrick, even if they are less showy.

One afternoon I pick up my phone and call Ira van Eelen. If anyone has a personal stake in the future of cultured meat, certainly it's the daughter of Willem van Eelen.

It's not as though, when her father died, Ira immediately picked up where he left off, instantly taking up the mantle of making cell-cultured meat a reality—pushing and jostling the science forward with his same brand of fervor. In fact, she turned her attention to other things for a while. She got married, raised a family, built a life on a beautiful houseboat, and dedicated herself to a career working in public relations with the public health sphere—dentistry, in particular.

When Tetrick first sought to acquire her father's patents, he called her to learn more about van Eelen and his work.

"That phone call from Josh really changed things for me," she tells me. Tetrick set her on a different path.

In the following two years, she focused her time and energy on promoting cultured meat, pushing her father's legacy closer and closer to the finish line. The work has grown into an obsession. In many ways, she says it's made her life demonstrably tougher to navigate—a statement to which I'm sure her father would give a knowing smile.

Since Tetrick first reached out to her, Ira started the Alternative Protein Show, a conference that seeks to bring like-minded food technologists together. She's spent countless hours trying to engage with farmers in the Netherlands, working to convince them that there is a role for them in a reenvisioned cell-cultured meat future. She says she's given presentations to the European Commission and the Christian Democratic Appeal party in the Netherlands, an attempt to educate policymakers about cultured meat and why it would be a positive thing for the country to embrace.

So much of what she's doing boils down to communication— to taking her own experience and knowledge and finding an effective way to talk to people about it without scaring them.

She's had her ups and downs with Tetrick, particularly after the Netherlands deal went bust. She calls herself a straight shooter, and admits that some of her phone calls with Tetrick around strategy have been tense at times.

"Josh is not always a nice person, but he has willpower, and imagination, and a story to tell," she says. "You need special people who have imagination, and stamina, and willpower."

She's not wrong. People may be more interested in trying new versions of meat than ever, but the numbers are daunting. Some outlooks anticipate the global meat sector will be valued at $1.14 trillion by 2023. That means, if estimates by folks in the industry are accurate, that alternative options will have to be valued at about $57 billion to make even a chip in the global market.

We're in an incredibly important moment in human history with a lot at stake. So much of the success of this particular movement boils down to storytelling—to distilling science and experience and knowledge and finding a way to connect it to people's lives and concerns without frightening them. Perhaps, on some level, Tetrick is precisely what the moment needs: a good storyteller. And maybe that story will begin in Israel. Or maybe it will begin in the United States. Or Singapore.

Two years earlier, when I first stepped into JUST, Tetrick told me about his improbable life story. We talked about how much Balk had inspired and directed him at crucial steps along the way, how he felt so untethered to a specific path, how even in times of stability he had an urge to shake things into chaotic motion again.

"Not to compare myself to a guy I'm about to mention right now—although it's going to sound like this—but I've been reading this biography of Ulysses S. Grant," he began.

I looked into Tetrick's face skeptically. Was he really about to compare himself to a celebrated general in the American Civil War? He smiled and continued to tell his story, about how when Grant was in his thirties he worked at his father's leather store in Galena, Illinois, bored with the day-in-day-out banalities of being a shopkeeper. He didn't like dealing with customers or learning the ins and outs of running a store.

"He was thirty years old and lost, right?" Tetrick said. "And then the war started."

He let me connect the dots: a man unsure of what to do with his life, somehow rising to the occasion. When I first started talking to Tetrick, he was hoping he'd have cell-cultured meat somewhere in the market by the beginning of 2018. That didn't work. Then, for months, he confidently told me he'd have it somewhere by the end of the year. Since then, he's hopscotched from Amsterdam, to the United Arab Emirates, to Hong Kong, Singapore, and finally to the end of 2019. Still nothing.

I do believe he's close. I think most of the companies in this field are close, as well—even ready right now, but stymied by the money and power and sheer inertia holding the status quo in place.

Tetrick is facing our challenges head-on, from animal welfare to climate change. The moment requires good science, good storytelling, and action.

"It's game time," Tetrick says.

Maybe he'll rise.

ACKNOWLEDGMENTS

We live in a wild and wayward world—this has always been the case. But in our particular modern moment, that world is punctuated by a climate crisis that is at once difficult to wrap our heads around and seemingly unavoidable.

When I was younger, I was told by people in older generations that the changing climate was the product of Mother Earth's own natural rhythms. There were ice ages before, and there would be ice ages again—humans were insignificant players in a natural world that dictates our fate, not the other way around. It's a simple and convenient logic, one that was easy to embrace because, in spite of a mountain of complex science and decades of warning signs, it was less scary than the alternative.

The data continue to prove those people wrong; meanwhile, the world is shaping up to be a more dangerous place for their children. The quality of our soil is degrading. Agricultural lands are being recast by fire and drought. Increased hurricane activity and flooding threaten the infrastructures of coastal cities.

And by 2050, it is expected that Latin America, sub-Saharan Africa, and Southeast Asia will generate some 143 million climate migrants—refugees cast from their homes in an increasingly volatile world.

And then the data hit home. In 2017, one of my closest friends informed me, after consideration, that she and her husband had decided it would be unethical for them to have children, to bring new life into a world beset by this upcoming crisis. Then another couple told me the same thing. And then another. I wondered to myself then—and I still wonder now—when was the last time in human history that a generation questioned the ethics of procreation because mankind had failed in its duty to be a good steward of the planet?

The idea to write a book on cell-cultured meat came about through sheer happenstance because my wonderful editor, Merry Sun, reached out and challenged me to think about tackling a topic that's truly meaningful. Without her prodding, I'd never have considered taking a deeper dive into an innovation with so much promise for our world. Every bit of this book she touched made it a better read.

Thanks to my agent, Peter Steinberg, for helping me craft a compelling proposal for a story that is, at its core, all about creating the change you want to see in the world. Many of us aspire to foster positive change on the planet—to carry out our fundamental duty to repair the world—but it's almost never easy to navigate just how exactly to do that. This book only touches on the interesting topic of how people in the vegan movement evolved, updating how they interact with and even subvert monolithic systems to try and create positive change. One of those people is the central figure in this book. Thank you to Josh Tetrick for providing a window into this space and for being willing to answer tough questions. Thank you also to Ira van Eelen, who opened her home to me on two occasions as I sought to learn as much as I could about her father, Willem van Eelen, who fought—perhaps harder than anyone—to make cell-cultured meat a reality.

To the hardworking people at New America, thank you for your support, which proved invaluable during the reporting process.

Much gratitude to Brad Lovett, who supported me through every step of the reporting and writing process, and whose own enthusiasm about this topic inspired me when I needed the encouragement most. I also want to thank my oldest friend, Erin Palmer, along with John Palmer (I'm sorry the upside-down pig idea didn't work), Helena Bottemiller Evich, Yotam Shwimmer, and Roei Bar Cohen for their thoughtful and invaluable first readings of the manuscript. Our conversations grounded me.

Not to be forgotten are the friends who kept me sane as I navigated my way through reporting and writing. Special shoutouts to Fareed Choudhry, Andrew Dewan, and Scott Kramer, for providing distracting fun on snowy nights; Eamonn Wright for the martini chats; Zach Cohen for his shooping and positive energy; and also to Danny Volker, Spencer Kornhaber, Jane Hu, Cristobal Valencia, Armen Shahnazarian, Rose Reid, Emily Lamb, Lauren Chester White, Alex White, Benji Jones, Lindsey Albracht, and Bryan Milo.

ACKNOWLEDGMENTS

Finally—and most importantly—I want to thank my parents, Bruce and Connie; my three siblings, Luke, Abraham, and Livia; and my nana, Carolyn.

In particular, thank you to my mother, who, during the course of homeschooling me through my early education, always encouraged my writing. Without that support, I doubt I'd ever have found my way into journalism, let alone write a book. She was my first in a line of influential editors, a thoughtful fellow food nerd, and the kindest person I know.

NOTES

Prologue

xiii   *make-or-break moment:* Kate O'Riordan, Aristea Fotopoulou, and Neil Stephens, "The First Bite: Imaginaries, Promotional Publics and the Laboratory Grown Burger," *Public Understanding of Science* 26, no. 2 (August 2, 2016): 148–63. https://doi.org/10.1177/096366251 6639001.

Chapter One: Digging In

3   *often grow better:* Paul Shapiro, "Chicken Might Be the First Lab-Grown Meat to Make It to Your Grocery Store," *Vice,* January 2, 2018. www.vice.com/en_us/article/3k5ak3/chicken-might-be-the-first-lab -grown-meat-to-make-it-to-your-grocery-store.

6   *factoring in deforestation:* "Special Report: Climate Change and Land," IPCC, 2019. www.ipcc.ch/2019/08/08/land-is-a-critical-re source_srccl/.

6   *cow manure:* Henning Steinfeld, *Livestock's Long Shadow: Environmental Issues and Options* (Rome: Food and Agriculture Organization of the United Nations, 2006).

6   *nuanced findings:* P. J. Gerber, et al., *Tackling Climate Change Through Livestock: A Global Assessment of Emissions and Mitigation Opportunities* (Rome: Food and Agriculture Organization of the United Nations, 2013).

6   *about 8 percent:* Ibid.

7   *heat-trapping power as carbon:* G. Yvon-Durocher, et al., "Methane Fluxes Show Consistent Temperature Dependence Across Microbial to Ecosystem Scales," *Nature* 507 (2014): 488–91. doi:10.1038/nature13164.

7   *100 kilograms of methane:* Fred Pearce, "Grass-Fed Beef Is Bad for the Planet and Causes Climate Change," *New Scientist*, October 3, 2017. www.newscientist.com/article/2149220-grass-fed-beef-is-bad-for-the-planet-and-causes-climate-change/.

7   *produce 1 pound of beef:* Tamar Haspel, "Vegetarian or Omnivore: The Environmental Implications of Diet," *The Washington Post*, March 10, 2014. www.washingtonpost.com/lifestyle/food/vegetarian-or-omnivore-the-environmental-implications-of-diet/2014/03/10/648fdbe8-a495-11e3-a5fa-55f0c77bf39c_story.html.

8   *fewer than 2.06 million:* "Farming and Farm Income," U.S. Department of Agriculture, November 27, 2019. www.ers.usda.gov/data-products/ag-and-food-statistics-charting-the-essentials/farming-and-farm-income/.

8   *dairy herd shrank:* Jim Dickrell, "Licensed Dairy Farm Numbers Drop to Just Over 40,000," *Milk Business: Farm Journal & MILK Magazine*, February 21, 2018. www.milkbusiness.com/article/licensed-dairy-farm-numbers-drop-to-just-over-40000.

8   *1 trillion fish a year:* Alison Mood, "Worse Things Happen at Sea: The Welfare of Wild-Caught Fish," Fish Count, 2010. www.fishcount.org.uk/published/standard/fishcountfullrptSR.pdf.

8   *relied on slave labor:* Margie Mason, "Fishing Slaves No More, but Freedom Brings New Struggles," Associated Press, July 12, 2017. www.ap.org/explore/seafood-from-slaves/fishing-slaves-no-more-but-freedom-brings-new-struggles.html.

12   *next evolution:* Isha Datar, "Why Cellular Agriculture Is the Next Revolution in Food," *Food Tech Connect*, January 7, 2019. https://foodtechconnect.com/2016/04/11/cellular-agriculture-is-the-next-revolution-in-food/.

14   *wrote an article:* Winston Churchill, "Fifty Years Hence," *Popular Mechanics*, March 1932. http://rolandanderson.se/Winston_Churchill/Fifty_Years_Hence.php.

14   *2011 Oxford University study:* Hanna L. Tuomisto and M. Joost Teixeira de Mattos, "Environmental Impacts of Cultured Meat Production," *Environmental Science & Technology* 45/14 (2011): 6117–23. doi: 10.1021/es200130u.

15 *play a role:* Wyatt Bechtel, "Cattlemen's Groups Voice Concerns with Lab-grown Meat to USDA, FDA," *Drovers,* October 24, 2018. www .drovers.com/article/cattlemens-groups-voice-concerns-lab-grown -meat-usda-fda.

Chapter Two: A Culinary Godfather

17 *"vat stuff":* William Gibson, *Neuromancer* (New York: Ace Science Fiction Books, 1984).

17 *ChickieNob:* Margaret Atwood, *Oryx and Crake* (New York: Nan A. Talese, 2003).

24 *began to ring:* Chase Purdy, "The Idea for Lab-Grown Meat Was Born in a Prisoner-of-War Camp," *Quartz,* August 8, 2018. https://qz.com /1077183/the-idea-for-lab-grown-meat-was-born-in-a-prisoner-of -war-camp/.

25 *successfully culture:* "Morris Benjaminson, Laboratory Meat Pioneer—Obituary," *The Telegraph,* May 26, 2017. www.telegraph .co.uk/obituaries/2017/05/26/morris-benjaminson-laboratory-meat -pioneer-obituary/.

26 *small tribute:* Ingrid Newkirk, "The Finalists," *The New York Times,* May 6, 2012. https://archive.nytimes.com/query.nytimes.com/gst/full page-9802E2D9103BF935A35756C0A9649D8B63.html.

Chapter Three: The Molecular Miracle

33 *upended that theory:* "Life's Limit," *RadioLab,* WNYC Studios, June 14, 2007. www.wnycstudios.org/podcasts/radiolab/segments/91563-lifes -limit.

35 *control the growth:* Prestage Department of Poultry Science website, North Carolina State University, Paul Mozdziak's page. https://cals .ncsu.edu/prestage-department-of-poultry-science/people/pemozdzi/.

37 *hundreds of proteins:* H. Lodish, et al., *Molecular Cell Biology,* 4th ed. (New York: W. H. Freeman, 2000), Section 6.2, "Growth of Animal Cells in Culture." Available from: www.ncbi.nlm.nih.gov/books/NBK 21682/.

44 *"whole new industry":* Isha Datar, "The Future of Food Is Farming Cells, Not Cattle," *Quartz,* October 18, 2018. https://qz.com/1383641 /the-future-of-food-is-farming-cells-not-cattle/.

48 *"don't play at all"*: Purdy, "The Idea for Lab-Grown Meat Was Born in a Prisoner-of-War Camp."

Chapter Five: Panic in Amsterdam

59 *cookbook:* Koert van Mensvoort and Hendrik-Jan Grievink, *The In Vitro Meat Cookbook* (Amsterdam: BIS Publishers, 2014).

62 *"vegan currywurst":* James Kanter, "Take Feta. Add Frites. Stir in European Food Rules. Fight," *The New York Times,* June 22, 2017. www .nytimes.com/2017/06/21/business/eu-food.html.

Chapter Six: Untethered

73 *told* **The Guardian:** John Fecile, "'Banned in 46 Countries'—Is *Faces of Death* the Most Shocking Film Ever?" *The Guardian,* October 1, 2018. www.theguardian.com/film/2018/oct/01/banned-in-46-countries-is -faces-of-death-the-most-shocking-film-ever.

77 *penned an op-ed:* Josh Tetrick, "You Can Save the Planet," *Richmond Times-Dispatch,* March 15, 2009. www.richmond.com/news/you-can -save-the-planet/article_c0492079-ad99-5dbf-bfa0-d9d991d5b218 .html.

83 *Sancho Panza:* "Solina Chau," *Forbes.* Accessed December 4, 2019. www.forbes.com/profile/solina-chau/#c372e553b939.

Chapter Seven: The Art of War

88 *Andrew Zimmern:* Andrew Zimmern, "Stop Bullying Sustainable Food Companies," Change.org. Accessed December 4, 2019. www .change.org/p/tell-unilever-to-stop-bullying-sustainable-food -companies.

89 *Stephanie Strom:* Stephanie Strom, "Hellmann's Maker Sues Company over Its Just Mayo Substitute Mayonnaise," *The New York Times,* November 10, 2014. www.nytimes.com/2014/11/11/business/unilever -sues-a-start-up-over-mayonnaise-like-product.html.

90 *Joanne Ivy:* Ivy declined a request for comment.

92 *emails obtained:* Dan Charles, "How Big Egg Tried to Bring Down Little 'Mayo' (and Failed)," *The Salt,* National Public Radio, September 3, 2015. www.npr.org/sections/thesalt/2015/09/03/437213511/how -big-egg-tried-to-bring-down-little-mayo-and-failed.

Chapter Eight: The Lost Puppy

96 **Serious Eats** *food blog:* J. Kenji López-Alt, "Which Vegan Mayo Is the Best?" *Serious Eats,* August 10, 2018. www.seriouseats.com/2014/02 /vegan-mayonnaise-taste-test-produces-surprising-results.html.

97 **The Splendid Table:** Jack Bishop, "America's Test Kitchen Finds a Vegan Mayonnaise Game-Changer," *The Splendid Table*, March 2, 2017. www.splendidtable.org/story/americas-test-kitchen-finds-a-vegan -mayonnaise-game-changer.

101 *scathing review:* "'Dear Josh'—JUST," Glassdoor, September 28, 2017. www.glassdoor.co.in/Reviews/Employee-Review-JUST-RVW 17073899.htm.

101 *report by* **Bloomberg:** Olivia Zaleski, Peter Waldman, and Ellen Huet, "How Hampton Creek Sold Silicon Valley on a Fake-Mayo Miracle," *Bloomberg Businessweek*, September 22, 2016. www.bloomberg.com /features/2016-hampton-creek-just-mayo/.

105 *Series E:* Beth Kowitt, "Hampton Creek, Now a Unicorn, Shakes Up Management Team," *Fortune*, May 2, 2017. https://fortune.com/2017 /05/01/hampton-creek-unicorn-management/.

Chapter Nine: The Rest of the Herd

115 *Global Food Security Index:* "Global Food Security Index 2017," Economist Intelligence Unit, October 2017. https://foodsecurityindex.eiu .com/.

116 *agricultural production:* Associated Press, "UN: Farmers Must Produce 70% More Food by 2050 to Feed Population," *The Guardian*, November 28, 2011. www.theguardian.com/environment/2011/nov /28/un-farmers-produce-food-population.

119 *raised $17 million:* Paul Sawers, "Lab-Grown Food Startup Memphis Meats Raises $17 Million from DFJ, Cargill, Bill Gates, Others," *VentureBeat*, August 23, 2017. https://venturebeat.com/2017/08/23/lab -grown-meat-startup-memphis-meats-raises-17-million-from-dfj -cargill-bill-gates-richard-branson-others/.

119 *"broadening our exposure":* "Future Meat Technologies Raises $14 Million in Series A Funding, Announces Pilot Production Facility," press release, Future Meat Technologies, October 10, 2019. www.prnewswire .com/news-releases/future-meat-technologies-raises-14-million-in -series-a-funding-announces-pilot-production-facility-300936425 .html.

122 *first investment:* Cathy Siegner, "Merck's Venture Capital Arm Invests $8.8M in Mosa Meat," *Food Dive*, July 17, 2018. www.fooddive.com /news/mercks-venture-capital-arm-invests-88m-in-mosa-meat /527885/.

Chapter Ten: The Belly of the Beast

142 *political vandalism:* Laurie Goodstein, "N.C. Trial Conjures Up Anti-war Era," *The Washington Post,* February 15, 1994. www.washington post.com/archive/local/1994/02/15/nc-trial-conjures-up-antiwar -era/1b93347e-56d3-4e5f-9eba-7612a0655f5a/.

144 *John Dear:* John Dear, "Remembering the 20th Anniversary of a Plowshares Action for Peace," *National Catholic Reporter,* December 3, 2013. www.ncronline.org/blogs/road-peace/remembering-20th -anniversary-plowshares-action-peace.

148 *"VEGETARIANS AT THE GATE":* Thomas Buckley, "The Vegetarians at the Gate," *Bloomberg Businessweek,* December 19, 2018. www.bloomberg .com/news/features/2018-12-19/the-vegetarians-at-the-gate.

149 *global temperature rise:* Zoë Schlanger, "The UN All but Admits We Will Probably Pass the 1.5°C Point of No Return," *Quartz,* November 26, 2019. https://qz.com/1755954/un-climate-report-says-warming -past-1-5c-is-likely/.

150 *UN report said:* United Nations Environment Programme, *Emissions Gap Report 2019* (Nairobi: UNEP, 2019).

Chapter Eleven: Food Fight

155 *grind to death:* John MacDougall, "By 2020, Male Chicks May Avoid Death By Grinder," *National Geographic,* June 13, 2016. www.national geographic.com/culture/food/the-plate/2016/06/by-2020—male -chicks-could-avoid-death-by-grinder/.

155 *"great grandparents":* Jessica Almy, The Good Food Institute, April 17, 2018. www.gfi.org/images/uploads/2018/04/GFIetal-Comment-FSIS -2018-0016.pdf.

156 *congressional effort:* Chase Purdy, "Trump May Get the Last Word on the Longstanding Fight over Whether Almond Milk Is Actually 'Milk,'" *Quartz,* March 3, 2017. https://qz.com/923234/theres-a-war -over-the-definition-of-milk-between-dairy-farmers-and-food -startups-and-donald-trump-may-settle-it/.

158 *"swindle of the age":* John Suval, "W. D. Hoard and the Crusade Against the 'Oleo Fraud,'" *Wisconsin Historical Society,* 2012.

159 *Hoard's paranoia:* Ibid.

159 *twenty-six states convened:* Richard A. Ball and J. Robert Lilly, "The Menace of Margarine: The Rise and Fall of a Social Problem," *Social Problems* 29/5 (June 1, 1982): 488–98. https://doi.org/10.2307/800398.

159 *signed a bill into law:* Gerry Strey, "The 'Oleo Wars': Wisconsin's Fight over the Demon Spread," *Wisconsin Historical Society,* 2007.

163 *"Menace of Margarine":* Ball and Lilly, "The Menace of Margarine."

172 *taking a beating:* Emma Newberger, "'Trump Is Ruining Our Markets': Struggling Farmers Are Losing a Huge Customer to the Trade War—China," CNBC, August 13, 2019. www.cnbc.com/2019/08/10/trump-is -ruining-our-markets-farmers-lose-a-huge-customer-to-trade -war——china.html.

173 *told entrepreneurs:* Susan Mayne, "FDA's Role in Supporting Innova- tion in Food Technology," U.S. Food and Drug Administration, March 22, 2018. www.fda.gov/about-fda/what-we-do-cfsan/fdas-role-support ing-innovation-food-technology.

175 *letter of complaint:* Chase Purdy, "US Food Regulators Are Fighting over Who Gets to Oversee Cell-Cultured Meat," *Quartz,* July 13, 2018. https://qz.com/1327919/us-food-regulators-are-fighting-over-who -gets-to-oversee-cell-cultured-meat/.

176 *directly to the president:* Chase Purdy, "Donald Trump May Decide the Regulatory Future of Cell-Cultured Meats," *Quartz,* August 1, 2018. https://qz.com/1340868/the-us-meat-industry-is-asking-donald -trump-to-decide-how-cell-cultured-meat-is-regulated/.

Chapter Twelve: Promise Abroad

183 *Climate Central:* "Report: Flooded Future: Global Vulnerability to Sea Level Rise Worse than Previously Understood," Climate Cen- tral, October 29, 2019. www.climatecentral.org/news/report-flooded -future-global-vulnerability-to-sea-level-rise-worse-than-previously -understood.

183 *World Economic Forum:* Xi Hu, Environmental Change Institute, and University of Oxford, "Where Will Climate Change Impact China Most?" World Economic Forum, April 5, 2016. www.weforum.org /agenda/2016/04/where-will-climate-change-impact-china-most/.

184 *development of cell therapy:* Catherine Lamb, "Singapore to Invest $535 Million in R&D, Including Cultured Meat and Robots," *The Spoon,* March 29, 2019. https://thespoon.tech/singapore-to-invest -535-million-in-rd-including-cultured-meat-and-robots/.

Chapter Thirteen: The Ties That Bind

192 *Michael Pollan:* Michael Pollan, "An Animal's Place," *The New York Times*, November 10, 2002. www.nytimes.com/2002/11/10/magazine /an-animal-s-place.html.

196 *Peter Brabeck-Letmathe:* Chase Purdy, "'Nature Is Not Good to Human Beings': A Food Industry Titan Makes the Case for a New Kind of Diet," *Quartz*, December 27, 2016. https://qz.com/856541/the -worlds-biggest-food-company-makes-the-case-for-its-avant-garde -human-diet/.

199 *food yields: The State of Agricultural Commodity Markets 2018: Agricultural Trade, Climate Change and Food Security* (Rome: Food and Agriculture Organization of the United Nations, 2018).

200 *risks of what's coming:* "Agriculture and Climate Change Adaptation," California Department of Food and Agriculture, 2019. www.climate change.ca.gov/adaptation/agriculture.html.

Chapter Fourteen: Setting the Table

231 *global meat sector:* "Global Meat Sector Market Analysis & Forecast Report, 2019—A $1.14 Trillion Industry Opportunity by 2023," Globe Newswire News Room, Research & Markets, May 2, 2019. www.globenewswire .com/news-release/2019/05/02/1815144/0/en/Global-Meat-Sector -Market-Analysis-Forecast-Report-2019-A-1-14-Trillion-Industry -Opportunity-by-2023.html.

INDEX

INDEX

Printed in the United States
by Baker & Taylor Publisher Services